Numerical Modeling of Emissions and Thermoacoustics in Heavy-Duty Gas Turbine Combustion Systems

zur Erlangung des akademischen Grades eines

DOKTORS DER INGENIEURWISSENSCHAFTEN (Dr.-Ing.)

der Fakultät für Chemieingenieurwesen und Verfahrenstechnik des

Karlsruher Instituts für Technologie (KIT)

genehmigte

DISSERTATION

von

Stefan Dederichs M.Sc.

aus Düsseldorf

Referent: Prof. Dr.-Ing. Nikolaos Zarzalis
Korreferent: Prof. Dr.-Ing. Karlheinz Schaber
Tag der mündlichen Prüfung: 08.12.2016

Bibliografische Information der Deutschen Nationalbibliothek

Die Deutsche Nationalbibliothek verzeichnet diese Publikation in der
Deutschen Nationalbibliografie; detaillierte bibliografische Daten sind
im Internet über http://dnb.d-nb.de abrufbar.

ISBN 978-3-8325-4414-0

Logos Verlag Berlin GmbH
Comeniushof, Gubener Str. 47,
10243 Berlin
Tel.: +49 (0)30 42 85 10 90
Fax: +49 (0)30 42 85 10 92
INTERNET: http://www.logos-verlag.de

Preface

The underlying work was done at the division for combustion technology of the Engler-Bunte-Institute at the Karlsruhe Institute of Technology in cooperation with the Siemens AG.

I would like to express my appreciation to Professor Zarzalis who was my supervisor. He offered me the chance to do research work and he gave me a great insight into the teaching culture at his chair in Karlsruhe. But I also had the chance to find out about industrial work. Furthermore, I like to thank Professor Schaber for his interest and the willingness to act as the second reviewer. Dr. Christian Beck and Dr. Werner Krebs from the Siemens AG gave me the chance to work on-site in Mülheim (Germany) and Orlando (USA). They allowed me to participate in the development of a new prototype combustion system. This mix of industrial and scientific work was a great chance for me and I greatly appreciate the huge commitment of Professor Zarzalis, Christian and Werner.

I would like to thank DLR Stuttgart and Siemens AG for the experimental work. Furthermore, I like to thank Armin Wehrfritz, who supported the implementation of the mixture average approach. I am grateful to Dr. Oliver Lammel, Dr. Feichi Zhang, Dr. Vinayaka Nakul Prasad, Dr. Peter Habisreuther and Fabian Proch for many helpful discussions.

Personally I would like to thank my family and friends. They supported me with patience and sympathy over many years.

Parts of this work were performed on the computational resource bwUniCluster funded by the Ministry of Science, Research and the Arts and the Universities of the State of Baden-Württemberg, Germany, within the bwHPC framework program.

Düsseldorf, September 2016   Stefan Dederichs

Contents

Nomenclature

Upper case	Roman symbols	Unit
A	Area	$[m^2]$
A	pre-exponential factor	[variable]
C_P	Specific heat capacity at constant pressure	$[J/kg/K]$
C_S	Constant of the Smagorinsky model	$[-]$
C_V	Specific heat capacity at constant volume	$[J/kg/K]$
D	Diffusion coefficient	$[m^2/s]$
Da	Damköler number	$[-]$
E	Efficiency function	$[-]$
F	Thickening factor	$[-]$
Ka	Karlovitz number	$[1/m]$
L	Length	$[m]$
Le	Lewis number	$[-]$
M	Molar mass	$[kg/mol]$
Ma	Mach number	$[-]$
Pr	Prandtl number	$[-]$
Q	Heat	$[J]$
R	Universal gas constant	$[J/mol/K]$
Re	Reynolds number	$[-]$
Sc	Schmidt number	$[-]$
S	Flame speed	$[m/s]$
T	Temperature	$[°C \text{ or } K]$
V	Volume	$[m^3]$
V_k	Diffusion velocity	$[m/s]$
$\dot{V}$	Volume flow	$[m^3/s]$
X	Mole fraction	$[-]$

Y	Mass fraction	$[-]$
Z	Mixture fraction	$[-]$

Lower case	**Roman symbols**	**Unit**
a	Thermal diffusivity	$[m^2/s]$
c	Reaction progress variable	$[-]$
c	Speed of sound	$[m/s]$
d	Diameter	$[m]$
f	Frequency	$[Hz]$
h	Enthalpy	$[J/kg]$
k	rate coefficients	[variable]
$\dot{m}$	Mass flow	$[kg/s]$
p	Pressure	$[Pa$ or $bar]$
q	Specific heat	$[J/kg]$
r	Speed of reaction	$[mol/m^3 s]$
t	Time	$[s]$
u	Velocity	$[m/s]$
y^+	Dimensionless wall distance	$[-]$

	Greek symbols	**Unit**
β	Dimensionless heat loss parameter	$[-]$
ε	Turbulence dissipation rate	$[m^2/s^3]$
η	Kolmogorov length scale	$[m]$
Δ	LES filter width	$[m]$
λ	Acoustic wave length (c/f)	$[m]$
μ	Dynamic viscosity	$[kg/m/s]$
ν	Kinematic viscosity	$[m^2/s]$
$\nu_{i,k}$	Stichometric coefficients of the i-th element	$[-]$
Ξ	Wrinkling factor (S_{SGS}/S_L)	$[-]$
Π_{ij}	Viscous stress tensor	$[N/m^2]$
ρ	Density	$[kg/m^3]$
τ	Time lag, time-scale	$[s]$
φ	Arbitrary quantity	$[-]$
Φ	Equivalence ratio	$[-]$
$\dot{\omega}_k$	Chemical source term of species k	$[kg/m^3/s]$
Ω	LES filtering operator	$[-]$

Mathematical operators

x'	Temporal fluctuation
$\langle x \rangle$	Time averaged value
$\bar{x}$	Spatial filter
$\tilde{x}$	Favre weighted filter
$|x|$	Magnitude
x^*	Multiplied with constant factor for confidentiality

Abbreviations

1D	One-dimensional
3D	Three-dimensional
ATF	Artificially thickened flame
CFD	Computational Fluid Dynamics
CFL	Courant-Friedrichs-Lewy number
char	characteristic
DNS	Direct Numerical Simulation
eff	effective
ff, FF	flame front
FGM	Flamelet generated manifolds
FPLF	freely propagating (premixed) laminar flame
FSD	Flame surface density
gov	governing
GT	Gas turbine
JIC	Jet in crossflow
JPDF	Joint probability density function
K	Kolmogorov
l	laminar
LES	Large Eddy Simulation
PDF	Probability density function
pf, PF	post flame
PIV	Particle Image Velocimetry
PZ	Primary zone
RANS	Reynolds Averaged Navicr Stokes
SGS	Sub grid scale
SIMPLE	Semi-implicit method for pressure-linked equation
t	turbulent
th	thermal
TKE	Turbulent kinetic energy
TSS	Time scale separation
TTP	Transport and thermodynamic properties
TVD	Total variation diminishing
u	unburnt
VG	Vortex Generator

Chapter 1

Introduction

1.1 Motivation

From a technical point of view, today's key market requirements for the design of modern gas turbine (GT) combustion systems are fuel efficiency, operational flexibility and compliance with more restrictive emissions legislation, especially in terms of NO_x and CO emissions. This in particular means that emissions of CO and NO_x have to be below acceptable limits over a wide load range.

From an environmental perspective, NO_x, CO and CO_2 emissions have to be limited for the following reasons. NO_x causes photochemical smog, acid rain and destruction of ozone in the stratosphere [1]. CO emissions are toxic to humans [2] and influence global warming because they affect greenhouse gases [3]. An increased amount of CO_2 in the atmosphere is considered as the main reason for potentially irreversible global warming [4]. Therefore, emissions reduction and efficiency improvement are addressed by improved combustion technologies.

Lean premixed combustion has been established as a state-of-the-art technology for heavy-duty gas turbines. This kind of combustion process offers the advantage of low NO_x pollutant emissions [5]. However, lean premixed combustion is also associated with thermoacoustic instabilities [6]. Although the quantitative prediction of thermoacoustic instabilities remains challenging [7] it is important to predict them because they can affect the structural

1

integrity of a gas turbine.

With higher firing temperatures, NO_x emissions limit the operational range but increasing the firing temperature is driven by an efficient usage of fuel and the desire to reduce CO_2. On the other hand, it is important to extend the CO emissions compliant load range towards lower load conditions, i.e. lower combustion temperatures. These trends are boosted by an increasing contribution of renewables to power generation.

NO_x emissions increase exponentially with increasing flame temperature in lean premixed systems, and linearly with increasing residence time [8]. Instead, CO emissions correlate exponentially with decreasing temperature and residence time due to incomplete burn-out. Therefore, a tradeoff between CO and NO_x emissions regarding the residence time determines the combustor volume. The exponential dependence of emissions on the flame temperature raises a second design challenge, which is the quality (i.e. homogeneity) of the fuel-air mixture. Because the flame temperature is dependent on the fuel-air mixture, rich pockets of fuel-air mixture may drive NO_x emissions. In a technically non-perfectly mixed system, fuel-rich or lean pockets in the combustor may significantly contribute to increased total emissions compared to a perfect mixture. This has to be evaluated considering the constraint of thermoacoustic behaviors, because the fuel air mixing process is known to control some thermoacoustic instabilities [6].

The development of advanced gas turbine combustors requires a modeling capability for prediction early in the design phase. Beside the prediction of flow field and temperature distribution, the modeling of acoustics and emissions has become more important. There is an increasing trend towards higher gas turbine efficiency, and simultaneous emissions regulations continue to become stricter; this in consequence requires improved accuracy of the prediction during the design phase. Therefore, a high fidelity computational approach is proposed in this work, which can address the described multiple physical problems: the prediction of NO_x and CO emissions, taking thermoacoustics into consideration.

With increased computational resources, detailed simulations of a whole combustion system have become feasible [7, 9, 10]. These simulations rely on computational fluid dynamics (CFD) and are used in this work as a high-fidelity approach for the prediction of emissions and thermoacoustics. CFD simulations can be distinguished in steady-state (e.g., Reynolds averaged Navier-Stokes equations (RANS)) and transient calculations (e.g. large eddy simulation (LES)) [11]. Due to the pressure oscillating nature of the thermoacoustic effects, compressible LES has been established as the preferred method.

Moreover, the design process can involve optimization. This can be done by an incremental parametric variation of design features, but a typical optimization relies on a huge number of variants. Therefore, a further objective of this work is to present a second method to assess the NO_x emissions with reduced computational costs by means of a low order approach. The main input parameters of this approach are global firing temperature, premixing quality and residence time. This low order approach enables optimization runs to be performed with the premixing quality as a target quantity. The low order approach also helps during the early design phase to determine the combustor residence time.

1.2 Objective and outline of the thesis

The present work aims at providing improved prediction capabilities for key performance parameters of GT combustion systems, namely emissions and thermoacoustics. This leads to recommendations for the design process at the end of the work. But on the way to establish an optimized design process, different questions from a research point of view arose and have been answered. This thesis should help to support an ongoing tool development process working towards the replacement of experiments with computational simulations.

One precise objective is to demonstrate the feasibility of a compressible, tabulated chemistry-based LES approach for combined emissions and thermoacoustic prediction. Therefore, gaps in already published approaches for the prediction of emissions and thermoacoustics have to be identified and closed. Also an association between thermoacoustic effects and emissions should be addressed. Finally, the importance of transient effects in the context of NO_x emissions should be clarified.

Another objective of this thesis is to improve the design process of a modern GT combustion system. The design process relies on different methods, depending on its maturity. Such methods are described in the method Chapter 4 and Chapter 5. A distinction is made between a high-fidelity approach for a late design stage and a low order approach for the early design phase and optimization runs. After the method description, the validation of the novel approaches is presented.

After the Fundamentals in Chapter 2, Chapter 3 gives an overview of the state of the art methods. Both new introduced methods – the low order approach and the CFD-based approach rely on tabulated chemistry to gain high computational efficiency. Furthermore, both methods are based on the consideration of transient effects. The necessity of including transient effects is demonstrated in the full-scale validation in Chapter 7.

For the validation of the high-fidelity CFD approach, the complexity of the validation cases was increased stepwise, up to the technical case where emissions and self-excited thermoacoustic instabilities were predicted in a GT-scale combustion system. Part- and base-load conditions have been considered for different fuel staging schemes. The low order approach was also applied to the same case for a NO_x evaluation, while its validation was performed on a perfectly premixed lab-scale case. The lab-scale validation of the novel methods is shown in Chapter 6.

At the end of the work, Chapter 8 summarizes the research questions and answers which are addressed within the work. Finally, it is concluded that applying the proposed methods helps to promote the whole development chain from the very beginning, continuing to detailed optimizations at the end of the design process.

Chapter 2

Fundamentals

2.1 Laminar premixed flames

Laminar flames are of huge importance for the study of combustion phenomena. Simplicity allows to study them with moderate effort from numerical as well as from experimental point of view. It is a common procedure to decompose complex flames, like a turbulent flame, into laminar sub systems which than can be studied separately.

Several types of laminar flames are available which can be distinguished in perfectly premixed and diffusion type flames. Furthermore, for both type of laminar flames stretched and unstretched structures can be considered. Within the present work the perfectly premixed and unstreched type of laminar flames are investigated because they may represent a subsystem of a lean turbulent flame out of a GT combustion system. The degree of stoichiometry of a mixture can be described by means of the mixture fraction, Eq. (2.1). The mixture fraction is based on unburnt (indicator 'u') quantities, represented by the element mass fractions of fuel and oxidizer. The element mass fractions remain unaffected by a reaction because they represent the elements of the fuel and oxidizer molecules. The mixture fraction can be transferred into the equivalence ratio (ϕ) as shown in Eq. (2.2). An equivalence ratio below one describes a lean mixture, characterized by an excess of the oxidizer (e.g. air). Both quantities (ϕ and Z) are useful for modeling techniques because they are conservative due to the conservation of elements during combustion.

$$Z = \frac{Y_{fuel,u}}{\left(Y_{fuel,u} + Y_{oxidizer,u}\right)} \tag{2.1}$$

$$\phi = \frac{Y_{fuel,u}/Y_{oxidizer,u}}{(Y_{fuel,u}/Y_{oxidizer,u})\big|_{stoich}} = \frac{Z(1 - Z|_{stoich})}{(1 - Z)\, Z|_{stoich}} \tag{2.2}$$

A common model for the representation of a perfectly premixed, unstreched flame is the 1D, freely propagating, perfectly premixed, laminar flame (1D-FPLF) model. As exemplarily shown in Fig. 2.1 the whole 1D-FPLF can be divided into different zones. The present work distinguishes between the flame-front and the post-flame. The post-flame is characterized by the CO-oxidation. However, a so called reaction-zone [12] has been established before in order to explain the stabilization mechanism of the flame. For the stabilization, heat is transported from the reaction-zone towards the preheat-zone. Within the preheat-zone the mixture gets preheated until the auto-ignition temperature. The stabilization process of the flame relies on the equilibrium between the reaction and the molecular transport.

A characteristic quantity to describe laminar flames is the laminar flame speed (S_l [m/s]). At stationary conditions, this quantity is the velocity of the unburnt, fresh gas flow, coming from the left in Fig. 2.1. The laminar flame speed can be derived from the molecular quantities based on a dimensional analysis according to Buckingham as shown by Schmid [13]. The relevant molecular quantities are the thermal diffusivity a [m^2/s] and a global time scale of the reaction in a laminar environment τ_l [s]. The thermal diffusivity a [m^2/s] can be replaced by the diffusion coefficient D [m^2/s] or the kinematic viscosity v [m^2/s] by using the dimensionless numbers: Schmidt number ($Sc = v/D$) and Prandtl number ($Pr = v/a$). By means of a dimensional analysis of S_l, a and τ_l it follows Eq. (2.3).

$$S_l^2 \propto \frac{a}{\tau_l} \tag{2.3}$$

To model the 1D-FPLF simplifications can be made. For stationary conditions the derivation with respect to time is zero ($\frac{\partial}{\partial t} = 0$). Eq. (2.4) shows the resulting mass balance. Based on the derivation with respect to x only, the simplification for an one dimensional flow can be seen. This is appropriate because flat flames are expected based on experimental results [12]. The species transport equation Eq. (2.5) points out the equilibrium between

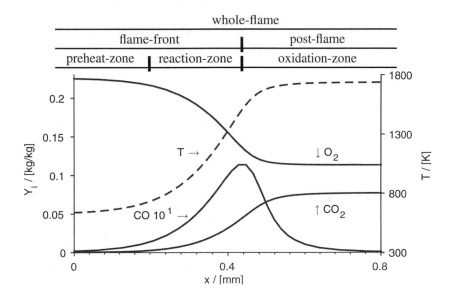

Figure 2.1: Species and temperature distribution of a freely propagating, laminar premixed flame for $\phi = 0.5$, $p = 10\ bar$ and methane as fuel.

convective transport on the left hand side and the diffusive transport plus the reaction (i.e. the chemical source term) on the right hand side. The equation makes use of the Fick's law for the diffusion velocity ($V_k = -D_k \frac{\partial Y_k}{\partial x}$). The source term ($\dot{\omega}_k$) can be closed by the Arrhenius law for the reaction rate.

- **Mass balance**

$$\frac{\partial}{\partial x}(\rho u) = 0 \tag{2.4}$$

- **Species transport**

$$\rho u \frac{\partial Y_k}{\partial x} = \frac{\partial}{\partial x}\left(\rho D_k \frac{\partial Y_k}{\partial x}\right) + \dot{\omega}_k \tag{2.5}$$

7

- **Energy balance**

$$\rho C_P\, u \frac{\partial T}{\partial x} = \frac{\partial}{\partial x}\left(\lambda \frac{\partial T}{\partial x} \right) - \frac{\partial T}{\partial x}\left(\rho \sum_{k=1}^{N} C_{P,k} Y_k V_k \right) - \sum_{k=1}^{n} h_k \dot{\omega}_k \quad (2.6)$$

- **Thermal equation of state**

$$p = \rho \frac{RT}{M} \quad (2.7)$$

The theory of molecules has been studied by Hirschfelder et al. [14]. They and others [15, 16] derived a model for the diffusion of a species k into the rest of a mixture based on the binary diffusion coefficients $D_{j,k}$ according to Eq. (2.8).

$$D_k = \frac{1 - Y_k}{\sum_{j \neq k} X_j / D_{j,k}} \quad (2.8)$$

A full set of elementary reactions (the so called chemical mechanism) is the decomposition of a global reaction. The chemical source terms Eq. (2.9) of the elementary reactions depend on their speed of reactions (r_i) and the stoichometric coefficients ($v_{k,i}$). The Arrhenius law Eq. (2.10) describes the rate coefficients k_{fw} and k_{bw} for the forward and backward reactions as the determining coefficients for the speed of a reaction. The Arrhenius law points out the exponential dependency of a reaction rate from the temperature and the activation energy (E_A). The pre-exponential factor A may also be temperature dependent but is approximated as constant in the below formulation as in the original Arrhenius law.

$$\dot{\omega}_k = \sum_{i=1}^{n} v_{k,i} r_i \left(k_{fw}, k_{bw} \right) \quad (2.9)$$

$$k = A \cdot exp(-E_A/RT) \quad (2.10)$$

To complete the description of the equations, needed for the solution of the 1D-FPLF, the energy equation Eq. (2.6) and the thermal equation of state Eq. (2.7) have to be mentioned. The energy equation describes the heat release of the global reaction as well as the temperature profile. The equation of state is needed to describe the relation between the density ρ, the pressure p and the temperature T.

A momentum equation is not needed for the description of the laminar 1D-FPLF but will be of relevance for turbulent flames in order to account for the three-dimensional (3D) expansion of the exhaust gases.

2.2 Turbulent flames

Turbulent flames are characterized by a turbulent flow environment. A turbulent environment can be observed if random variations in time and space take place. A conventional example for turbulent eddies can be seen in the smoke of a cigarette after a certain distance from the cigarette. Before the smoke becomes turbulent a transition from a regular laminar flow to the turbulent flow takes place. To characterize a turbulent flow regime, the Reynolds number can be used.

The Reynolds number is a number, used in the similitude theory. The number describes the ratio between inertial forces and counteracting viscous forces as shown in Eq. (2.11). The flow behavior of a duct flow for example becomes turbulent at a Reynolds number of about 2300. This number depends on the case, e.g. a free shear flow, as the flow of cigarette smoke, becomes turbulent at another Re-number.

$$Re = \frac{uL}{v} \tag{2.11}$$

The decay of turbulent scales with lifetime has been described by an energy cascade [17]. Turbulence is distributed over a wide range of length scales. While the production of turbulent eddies is associated with the big scales, the dissipation of the turbulent energy takes place at the smallest scales. At those small scales, kinetic energy can be converted into heat by the viscous forces. The size of the smallest turbulent scales has been described by Kolmogorov [18] based on a similitude model. Eq. (2.12) shows the definition of the Kolmogorov length scale as a function of the turbulence dissipation rate (ε). The Kolmogorov length scale as well as the Kolmogorov time scale, Eq. (2.13), are an indicator for the dissipation range in the energy cascade.

$$\eta_K = \left(\frac{v^3}{\varepsilon}\right)^{1/4} \tag{2.12}$$

$$\tau_K = \left(\frac{\nu}{\varepsilon}\right)^{1/2} \tag{2.13}$$

By using the turbulent Reynolds number (Re_t), the relation between the Kolmogorov micro scales and the macro scale of the flow can be described. Eq. (2.14) shows the definition of Re_t as the macroscopic velocity fluctuation (u') times the macroscopic turbulent length scale (L_t) over the microscopic kinematic viscosity (ν).

$$Re_t = \frac{u' L_t}{\nu} \tag{2.14}$$

Borghi [19] made use of the similitude theory to generate a classification of different flame regimes. Fig. 2.2 shows the log-log graph of Borghi with iso-lines for dimensionless quantities. On the x-axis the macroscopic turbulent length scale is normalized by the laminar flame thickness (δ_l). Accordingly, the macroscopic turbulent velocity fluctuation is normalized by the laminar flame speed on the y-axis.

$$\frac{u'}{S_l} = Re_t \cdot \left(\frac{L_t}{\delta_l}\right)^{-1} \tag{2.15}$$

$$\frac{u'}{S_l} = Da_t^{-1} \cdot \left(\frac{L_t}{\delta_l}\right) \tag{2.16}$$

$$\frac{u'}{S_l} = Ka^{2/3} \cdot \left(\frac{L_t}{\delta_l}\right)^{-1/3} \tag{2.17}$$

Eq. (2.15), Eq. (2.16) and Eq. (2.17) show how the iso-lines of the dimensionless quantities are obtained. The turbulent Damköhler number as defined in the Borghi diagram indicates the degree of wrinkling in a flame. Assuming a very low turbulent Damköhler number, the chemical resp. heat release time scale in a laminar environment (τ_l) would be longer than the turbulent macro time scale (τ_t). In this case, the turbulent eddies of all scales fit into the flame and the flame gets distributed due to the increase of turbulent diffusivity. The case is described by the distributed flame regime above the Da_t-line. The Karlovitz number is defined as the ratio between chemical resp. laminar time scale (τ_l) and the turbulent micro time scale (τ_K). The Karlovitz number is an indicator for stretch. The flamelet regime is located below a Karlovitz number of one. For this regime the chemical time scales are shorter than the shortest turbulent time scales and the flame front can be wrinkled by the full range of turbulent scales. According to Peters [20] the flamelet regime can be expected within a lean GT combustion system. Based on the turbulent Reynolds number it can be distinguished between laminar and turbulent flames.

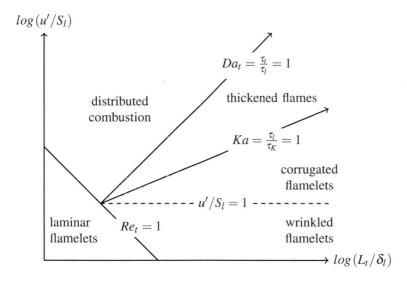

Figure 2.2: Regime diagram for premixed turbulent combustion according to Borghi [19]

To describe the fundamentals of a turbulent, 3D fluid dynamical problem, the Navier-Stokes equations can be solved. For reacting, compressible and 3D flows, an extended set of equations consist of *mass, species and energy balance equations* as well as the *equation of state* and a 3D *momentum equation*. In Eq. (2.18) the momentum equation is formulated incl. the time derivation term and incl. a sub grid scale (SGS) portion of the viscous stress tensor (Π_{ij}). Furthermore, body forces like gravity have been neglected within the present formulation. The time derivation is needed for the whole set of equations if transient effects are of interest. The SGS viscous stress tensor is needed for

11

turbulence modeling within Large-Eddy Simulations (LES) of transient nature. The SGS tensor ($\Pi_{ij,sgs}$) points out that turbulent fluctuations behave like viscous stresses (Π_{ij}).

- **Momentum equation**

$$\frac{\partial \bar{\rho} \tilde{u}_i}{\partial t} + \frac{\partial (\bar{\rho} \tilde{u}_i \tilde{u}_j)}{\partial x_i} + \frac{\partial \bar{p}}{\partial x_i} = \frac{\partial}{\partial x_i} \left(\overline{\Pi}_{ij} - \bar{\rho} \widetilde{\Pi}_{ij,sgs} \right) \qquad (2.18)$$

As mentioned in the previous paragraph, LES may be applied to simulate transient effects. Contrary to the Reynolds Averaged Navier Stokes (RANS) approach, the LES approach is used to model just the filtered range of the turbulent energy cascade at the small scales (aiming at the dissipation range) while the rest gets resolved.

The shown momentum equation relies on density weighted, Favre filtered quantities as exemplarily shown in Eq. (2.19) for the velocity. This type of averaging is applied to avoid unintended additional correlations due to density changes as e.g. present for reacting flows. Eq. (2.20) shows the decomposition of the velocity into the Favre filtered quantity and the SGS fluctuations. Further details on the modeling approach are shown in Chapter 4.

$$\tilde{u}_i = \frac{\overline{(\rho u_i)}}{\bar{\rho}} \qquad (2.19)$$

$$u_i = \tilde{u}_i + u'_{i,sgs} \qquad (2.20)$$

The Navier-Stokes equations can be simplified according to the problem but for combustion modeling most of the physics have to be considered and the simplification is limited. E.g. if transient effects play a minor role in combustion, RANS equations can be applied. Or, if Mach number effects can be neglected, a so-called low Mach number combustion model can be used. For the low Mach number model, density changes due to combustion are considered only in the momentum equation, which decouples pressure and density. A typical application range is for Mach numbers below 0.3 (i.e. the static pressure is close to the total pressure), but this approach compromises acoustical effects because sound waves are filtered out. As already mentioned in the introduction, the present work aims to predict multi-physical problems in combustion systems where transient acoustic effects are of importance. Consequently, a compressible, high Mach number, LES type combustion model is used within the present work for CFD simulations.

2.3 NO_x and CO Emissions

Oxides of nitrogen (NO & NO_2) are collectively named NO_x. These molecules as well as CO molecules are emitted due to combustion processes and can be studied by kinetically controlled chemical mechanisms. A comprehensive understanding of these mechanisms is required to develop emissions reduction strategies. Chemical mechanisms are the basis of most of the simplified laminar flame models as described in the previous sections. Laminar flame models help to study the effects of the mechanisms on the emissions production from a more global perspective. After an exemplary study, using such a laminar flame model, the fundamentals of the NO_x and CO mechanisms are explained within this section.

To illustrate the trade-off between NO_x and CO Emissions, a 1D-FPLF model can be applied to sweep the primary zone temperature of a GT combustor (i.e. sweep of mixture fraction). Fig. 2.3 shows the results of an idealized combustion system at perfectly premixed conditions by using the GRI (Gas Research Institute) mechanism 3.0 [21]. Based on the figure a remaining operating window in between certain emissions limits could be illustrated. An emissions limit for CO and NO_x of 10 $ppmv$ for example would span the remaining operating window between 1480 K and 1880 K. The steep left part of the CO curve is kinetically driven because the residence time is too short to achieve full burn out. The position of the steep left part can be shifted in real engines by fuel staging, as shown in Chapter 7. A gas turbine runs into this limit at very lean part load conditions. In this kinetically driven left side of the diagram, the amount of CO is increasing exponentially with residence time reduction. This exponential behavior can be derived from the pathway of the CO concentration in a flamelet (e.g. Fig. 4.3) since the amount of CO decreases exponentially with the spatial coordinate, which is proportional to the residence time. The right part of the CO curve in Fig. 2.3 illustrates the equilibrium state. The CO emissions based on an equilibrium state are not relevant to emissions compliance in modern GT systems [22]. At normal operating conditions complete burnout of CO can be achieved. In contrast, NO_x emissions are far from equilibrium and will be discussed in more detail in the next paragraphs.

Zeldovich described a mechanism of thermal NO formation in 1946 [23]; it is still fundamental to both detailed kinetic reaction mechanisms and to global NO formation equations. This mechanism was created after some es-

13

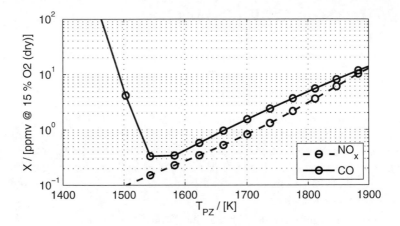

Figure 2.3: NO_x, CO trade off for perfectly premixed conditions as a function of the primary zone (PZ) temperature; The boundary conditions are a preheat temperature of 713 K, 20 bar pressure and 15 ms residence time.

sential simplifications were made, and is shown in Eq. (2.21). The first reaction relies on a relatively large activation energy, which makes it a kinetically limiting process. Consequently, this NO formation depends on the local temperature, which itself can be determined by the local mixture fraction, but other quantities such as pressure and fresh gas temperature also influence the formation.

$$N_2 + O \rightleftharpoons NO + N$$
$$N + O_2 \rightleftharpoons NO + O \qquad (2.21)$$
$$N + OH \rightleftharpoons NO + H$$

In the lean combustion regime, NO_x formation increases exponentially with the temperature due to the thermal NO formation pathway, as shown in Fig. 2.3. Therefore, premixing has been used successfully to reduce NO_x emissions by reducing the peak temperature of a lean flame [24,25]. Because of the exponential relation the net effect of poor mixing is an increase in NO_x emissions. Related to the trend towards lower NO_x emissions, the prediction of the influence of mixing quality on emissions is of huge relevance in the development process of a GT.

Perfect mixing in a GT combustion system usually cannot be achieved [26] for reasons such as limited premixing passage length, risk of flashback, auto ignition and pressure drop (mixing energy), but residence time also has to be considered because typical emissions limits are far below the equilibrium of NO_x. Therefore, reducing the volume of a combustion chamber results in lower NO_x emissions but is limited due to space resp. time needed for carbon oxidation. Due to the constant NO_x source term in the post-flame zone (see Fig. 5.2), NO_x emissions decrease linearly with a decreasing combustion chamber volume while CO increase exponentially. An additional source of high emissions might be thermoacoustic instabilities, which may cause mixture fluctuations which result in higher emissions.

Another well-known NO formation path was described by Fenimore [27]. He described the prompt NO formation (HCN initiation). Later, the N_2O formation path [28] was added to the set of NO_x formation pathways. The relevance of this path to leaner conditions in high-pressure combustion systems is discussed controversially [29]. The chemical pathways of thermal, prompt and N_2O based formations are covered by the GRI3.0 mechanism. Furthermore, an NNH path is covered [30]. Leonhard and Stegmaier [31] supported the theoretical work on NO_x emission formation in gas turbine combustors at perfectly premixed conditions with experimental data.

Contrary to the base load conditions, interesting for the NO_x emissions, part load conditions are the most challenging part of the CO prediction because the combustor volume might not provide enough space to reach equilibrium. CO is rapidly generated in the flame front, while the consumption is a rather slow process [32]. Michaud et al. [33] reduced the Miller and Bowman [34] mechanism to the most relevant reactions for lean CO combustion, Eq. (2.22), but also the equation $CO + O \rightleftharpoons CO_2$ should be considered to deduce the relation between CO and pressure, as follows. According to Le Chatelier's principle a chemical reaction tends to shift to the side of lower molecule density with increasing pressure. Therefore, the CO oxidation is accelerated at higher pressure conditions.

In Eq. (2.22) the OH radical influence is shown. Correa et al. [35, 36] described the radicals to be crucial for a quantitative prediction of CO. The radical oxidation is significantly faster than the CO oxidation due to hydrocarbons in the flame front [37]. Consequently, kinetically limited CO oxidation becomes relevant in the post-flame zone [32], where the reaction is rather slow.

$$HCO + M \rightleftharpoons H + CO + M$$
$$HCO + O_2 \rightleftharpoons HO_2 + CO$$
$$CO + OH \rightleftharpoons CO_2 + H \qquad (2.22)$$
$$HCO + OH \rightleftharpoons H_2O + CO$$

2.4 Thermo-acoustics

A famous thermo-acoustic phenomenon was described as early as 1859 by Rijke [38]. He described a standing wave in a tube, which was able to produce a sound. His experiment consisted of a vertically aligned open glass tube and a wired mesh which was attached inside the tube (at about 25 % of the length). The mesh acted as a heat source if it was heated right before the experiment by using a Bunsen burner aligned below the open end. Later, Lord Rayleigh [39] explained the mechanism responsible for the sound. The sound is created by a certain interaction between the heat release fluctuations and the pressure fluctuations. Such an interaction can be observed in GT combustion chambers as well and can be a source of thermo-acoustic instability.

Rayleigh defined a criterion [39] to determine whether acoustics are damped or amplified. Eq. (2.23) shows the requirement for an unstable state (i.e. amplification of the system). According to this definition amplification occurs if the pressure fluctuations p' and the heat release oscillations q' are in phase. This means in particular that a phase difference between those both quantities should be above $1/2\pi$ and bellow $3/2\pi$ to ensure stable operation. An extension of Rayleigh's criterion can be made for non-ideal conditions where acoustic energy can dissipate or be transmitted through a boundary. In the extended version, the amplification has to be above the losses in comparison to the original formulation, whereas a positive amplification is sufficient to reach an unstable situation. The effect of dissipation can be utilized to damp an acoustically unstable system (e.g. by using Helmholtz resonators [40]).

$$\int_V \int_0^\tau p'q' \, dt \, dV > 0 \qquad (2.23)$$

In unstable conditions, a so-called limit cycle is reached at the point where acoustic losses and acoustic gain are in equilibrium [41,42]. This equilibrium determines the pressure amplitude in an acoustically unstable state. Fig. 2.4 shows this equilibrium according to [41]. The figure illustrates that the non-linear behavior of the acoustic gain is necessary to reach equilibrium with the linearly increasing losses. These behaviors have been described by different research groups [43, 44]. To account for highly damped acoustic modes, which were according to Rayleigh's original formulation still untable, the wording stable refers in the present work to low relative pressure amplitudes below 0.5 %.

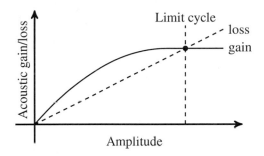

Figure 2.4: Thermoacoustic limit cycle description.

For the prediction of thermo-acoustic instabilities, low order (e.g. [45,46]) and different CFD-based approaches are established. The CFD-based approaches can be differentiated mainly into self-excited and forced approaches (e.g. [47]). Forced approaches rely on a separated analysis of the combustion acoustics in a low order code. For this approach, just the flame answer on perpetuated boundaries is simulated in a CFD environment [7]. Self-excited simulations cover the complete relevant physics with a fully compressible transient CFD method. The applicability of a self-excited approach to GT combustion systems has been demonstrated before [42, 48]. For example, Staffelbach et al. [48] have demonstrated the predictability of self-excited acoustic instabilities in a swirl type flame in an annular combustor. However, their approach was not based on tabulated chemistry. Furthermore, they were restricted to time steps corresponding to CFL numbers below one (named after Courant, Friedrichs and Lewy), while in the present work even the acoustic CFL number is above one.

$$CFL = \frac{|\mathbf{u}|\Delta t}{\Delta x} \qquad (2.24)$$

A novelty of the method used in the present work is that the combination of tabulated chemistry and flame front thickening has been used for the prediction of thermoacoustics. Furthermore, the computational costs were reduced in the present work by increasing the temporal discretization (i.e. increasing the time steps) which resulted in CFL numbers above one in some regions of the domain. The CFL-number, Eq. (2.24), is an indicator for the numerical stability as a function of the grid size and the time step width. The increased CFL number was possible due to an implicit time discretization. A more detailed investigation on an acoustic CFL number is shown in Chapter 4.

Chapter 3

State of the art

3.1 CFD based combustion modeling

Typical approaches to the modeling of combustion in a CFD environment rely on flame speed closure [49–52] or on significantly reduced reaction schemes [11]. The flame speed closure approach was presented in the 1970s by Zimont and summarized in a later work [49]. Hawkes and Cant [50] demonstrated the feasibility of the flame surface density (FSD) approach for transient Large Eddy Simulations (LES). Later, Pitsch [53] developed a G equation model for LES, which was formerly applied only for stationary simulations. However, more detailed mechanisms have also been used [9,54,55]. For such a detailed simulation, high computational resources are required to solve the underlying chemical kinetics. Therefore, a reasonable compromise is of interest, focused on the relevant physical effects without compromising the computational efficiency.

A state-of-the-art approach for transient phenomena is based on LES, where small fluctuations are filtered out. An introduction to the fundamentals of combustion LES was given by Janicka and Sadiki [56]. Without simplifying for low Mach numbers, the LES approach is appropriate for the simulation of thermo-acoustical phenomena [7, 11].

To model the kinetics of a combustion process, tabulated chemistry is an approach that requires modest additional computational costs without compromising fundamental information on pollutant chemistry. Different groups of researchers have investigated such approaches. For example, flamelet-generated manifolds (FGM) [57, 58], joint probability density function (JPDF) [59] and flame prolongation of intrinsic low-dimensional manifolds (FPI) [60, 61] are methods which are well established. These models are mainly based on the less expensive calculation of detailed 1D laminar paradigm flames (i.e., flamelets). Peters [20] has described the application range of the flamelet concept. This range is shown in the diagram of Borghi [19] in Fig. 2.2, where the flamelet regime is below a Karlovitz number of one. The flamelet regime is in particular located where the chemistry is faster than the turbulence because the chemical laminar time scales τ_l are smaller than the smallest turbulent Kolmogorov time scales τ_η. This is expected to be the case for lean GT combustion [20]. The thermochemical results of the laminar calculations are tabulated and mapped into a CFD calculation. As an extension, Kuenne et al. [62] have shown the FGM concept combined with artificial thickening of a flame (ATF) [63].

Tabulated chemistry relies on the calculation of a set of prototype flames in the range of interest. Those prototype flames are simplified to cover the main physical effects, e.g. a stirred-tank chemical reactor model with a continuous flow, which neglects diffusion but considers the effect of residence time. Other typical reactor types are a batch reactor, a burner stabilized flame, a freely propagating flat flame (1D-FPLF), a counter flow flame or a stagnation point flame. To represent a flamelet, 1D-FPLF have been established [61]. Those reactors are available in the commercial Software package *CHEMKIN* or in the open source package *Cantera* [64].The bases of all these simplified models are chemical kinetic mechanisms. They describe a set of Arrhenius type equations, which cover a set of species. Different mechanisms are dedicated to certain conditions. For example, the GRI (Gas Research Institute) mechanism 3.0 [21] is developed for natural gas type flames and its validity range covers GT-relevant conditions. This mechanism is optimized for laminar flame speed, ignition delay time and flame species concentration data [30]. However, also alternative mechanisms cover GT-relevant conditions: e.g. the Konnov 0.6 [65] and the GDF-Kin©3.0 [66]. The GRI3.0 mechanism has been chosen for this work because its computational efficiency is better than for the Konnov and the GDF mechanisms. The computational efficiency is better because the GRI mechanism is based on fewer species. The GRI mechanism is focused on lower hydrocarbons such as CH_4 while the other mechanisms also account for higher hydrocarbons, which are not of interest in this work. Typical NO_x and CO formations are considered by all mentioned mechanisms.

To account for the turbulence chemistry interaction, different methods are established, such as presumed PDF methods [67] or methods based on the description of the sub grid scale (SGS) flame wrinkling [68]. The disadvantage of presumed PDF methods in a tabulated chemistry context is the additional dimension which has to be added to the table to account for non-resolved fluctuations. The addition of table dimensions is costly computationally-wise because the look-up time increases. This approach seems to be more appropriate for RANS simulations than for LES because for LES equations, the effect of the modeled fluctuations is significantly smaller than for the RANS equations [69]. LES equations are partially resolved and the modeled fluctuations are limited by the filter scale. Even if the modeled LES fluctuations are rather small, they become important in the case of detailed chemistry modeling since they use detailed mechanisms. In this case, fast reactions, such as radical reactions, are simulated. Those reactions take place on very small scales. To overcome this challenge, a closed PDF approach, which is appropriate for LES, relies on stochastic fields [55] to characterize the influence of SGS fluctuations.

The combination of tabulated chemistry, artificial thickening and modeling of the SGS wrinkling seems to be a reasonable approach to simulate a GT combustion chamber. The usage of a tabulated chemistry approach is preferred for the previously mentioned reasons and since it enables modeling of pollutant emissions such as NO_x and CO through the addition of transport equations [70]. Thickening ensures that the flame is properly resolved and therefore the flame speed is predicted correctly. In combination with a flame wrinkling model, an accurate prediction of the flame position (i.e. turbulent flame speed) is expected. This is also of importance for the prediction of thermoacoustic phenomena. To the best of our knowledge, the combination of tabulated chemistry and artificial thickening has not been used before to predict thermoacoustic instabilities. This thesis helps to fill this gap.

3.2 Modeling of NO_x and CO Emissions

The prediction of NO_x and CO emissions in GT is still a challenging task because even minor changes in the local temperature may have a significant effect on kinetically-driven emissions. The CO emissions are particularly relevant for lean, low load operations. NO_x emissions are rather relevant for base load operations at higher firing temperatures. Because natural gas fired GT combustion systems are considered in this work, only the lean combustion of lower hydrocarbons is discussed. For lean combustion system mixing influences the local temperature and consequently the emission of both NO_x and CO pollutants.

CO Emissions

As previously mentioned in Chapter 2, Correa et al. [35, 36] emphasized the importance of detailed chemistry for CO prediction. Therefore, tabulated chemistry should be appropriate for the CO prediction because the influence of the radicals is considered. A remaining challenge for tabulated chemistry is the difference in time scales between the fast reaction in the flame front and the slow reaction in the post-flame region. Furthermore, the amount of molecules in the flame front is typically orders of magnitude higher than the amount of interest at the exit of a combustor.

Using CFD to resolve fast CO formation in the flame front by means of a transport equation is problematic on a finite grid due to the remaining cell length. Even if artificial thickening would help to describe the CO formation, it would be difficult to couple this to the governing species of the whole flame because the CO formation is not always representative of the whole flame. Another option would be to get the local amount of CO directly as a function of the whole flame governing species, but this also is not reasonable because of the different time scales between modeled whole flame and post flame. The effective time scale of CO_2, which is representative of the whole flame, on a typical CFD grid with about six cells over the flame front is fast in comparison to the slow CO oxidation. Even if the formation of CO_2 slows at the end of the reaction, this effect is not covered on a practical CFD grid because the small change of CO_2 at the end of the reaction is of the same order as the error induced by the discretization.

The challenges of using a separate transport equation for CO through tabulated chemistry for the source term closure have been addressed by Wegner et al. [70]. He introduced the time scale separation (TSS) model in a RANS environment. This model relies on an artificial source term for the net CO formation, while the slower CO oxidation in the post-flame zone is closed by a table of laminar source terms. Using tabulated chemistry requires a monotonic behavior of the governing species, which is actually not fulfilled by CO when considering the whole laminar pathway. However, if just the CO oxidation part is considered for the tabulation, this does result in a monotonic behavior of the CO source term as a function of the CO content itself. To complete the modeling approach, the artificially modeled source term of the CO formation is designed to provide the corresponding amount of CO at the position where the CO source term becomes zero.

NO_x Emissions

As already mentioned Chapter 2, the fuel-air mixing process is of major relevance for the NO_x pollutant generation. The effect of unmixedness on NO_x emissions was experimentally investigated by Fric [71]; he describes the NO_x emissions based on a spatial as well as a temporal unmixedness quantity. For the description of the mixture on a given plane, several definitions of the unmixedness are available [71, 72]. A typical definition is based on time-averaged data, which naturally neglects fluctuations in time. This deficit can be addressed by means of a spatiotemporal definition of the unmixedness. For a prediction of the impact of mixing effects, two approaches are established, namely detailed CFD simulations [58] and low order modeling [72].

Low order studies of NO_x emissions based on an integration of the detailed laminar chemistry over an assumed probability density function (PDF) of the equivalence ratio have highlighted the significance of unmixedness [73]. Another common alternative approach for the NO_x prediction is to model a network of different reactors [74], e.g. of perfectly stirred and plug flow reactors (PSR and PFR) to approximate a combustion chamber.

Network model approaches are of a generic nature and commonly do not consider mixing processes downstream of the flame front. Mixing processes have been investigated by means of experiments [75, 76] and simulations [77–79]. Models [80, 81] showed an exponential unmixedness decay over the residence time. One representative investigation of mixing was based on the jet-in-cross-flow phenomena [71]. Concepts for the combined consideration of mixing phenomena and detailed chemistry are a finite rate mixing model (FMM) [82] and a partially stirred reactor (PaSR) [83] model. These methods treat the fuel air mixing process in a similar way to a particle mixing process. The FMM model requires certain input parameters to describe the mixing process, which are not precisely known for a lean premixed GT combustor. Moreover, the computational effort is still high, due to the treatment of every particle as a batch reactor. In addition, these and the network-based approaches do not properly account for diffusive transport, which is of importance in the flamelet regime according to the Borghi diagram (Fig. 2.2).

A recent investigation has considered detailed kinetic mechanisms and laminar flamelets. Biagioli et al. [72] showed the effect of pressure and fuel-air unmixedness on NO_x emissions. They showed that the flame front NO_x emissions have a negative and the post-flame emissions have a positive pressure exponent [1]. They further introduced a two-zone model using a Damköhler criterion. These zones distinguish between NO_x formation in the flame front and the post-flame. A similar approach is used in this work for the low order approach.

Within the scope of a low order approach the computational effort for accurate NO_x prediction should be small to enable detailed parametric studies of design features. A pure CFD-based approach would not be computationally efficient in some cases, to evaluate e.g. a whole range of different mixing qualities, but it is necessary to get a mixing quality, which belongs to a certain target level of NO_x emissions.

The direct calculation of NO emissions within a CFD framework has been demonstrated before [58, 84]. Both publications used transient LES without modeling the rather small portion of NO_2. Furthermore, they both relied on a separate NO transport equation to overcome the slowly evolving post-flame zone. Ihme and Pitsch [84] investigated the radiation influence on NO, which is neglected in this work because its influence on the local temperature and emissions production is smaller for natural gas type flames in comparison to fuel oil based flames. The good predictability of NO emissions in a natural gas based combustion system without consideration of radiation was confirmed in the work of Ketelheun [58].

The present work also relies on a two-zone approach for the CFD-based NO prediction, but in contrast to the low order approach the post-flame region is defined via the net CO oxidation and starts at the position where the CO

---

[1]The exponent α was used in an expression of proportionality: $X_{NO_x} = X_{NO_x,0} \left(\frac{p}{p_0} \right)^{\alpha}$

source term is changing from positive to negative values. This approach is comparable to the time scale separation proposed by Wegner et al. [70] and allows for a combined prediction of CO and NO. In comparison to the low order approach, the CFD-based LES approach considers a complex flow field as well as relevant large scale temporal fluctuations. Such fluctuations may also be triggered by thermoacoustics.

Chapter 4

CFD based approach for emission & thermo-acoustic prediction

4.1 Introduction

As described in Chapter 1, this work is intended to improve the prediction of the key gas turbine performance parameters in order to improve the design process of a GT combustion system by developing new predictive methods. This chapter is dedicated to the high-fidelity method based on CFD. It aims to describe the simulation approach for the prediction of 3D unsteady aerodynamics, the combustion, the NO_x and CO pollutant formation and the thermoacoustics in a GT combustion system.

For the flame modeling, tabulated chemistry in combination with a novel formulation of a thickened flame approach is used. The emissions are predicted based on a two-zone approach which separates the CO production from the CO oxidation time scales. To ensure the predictability of acoustics with the CFD approach, a compressible solver with acceptable low numerical damping and a proper choice of boundary conditions have been used.

In the next section, the CFD solver, its fundamentals and the underlying numerical schemes are described. Thereafter, the combustion model formulation is presented. Within the section on the combustion model, the chemistry tabulation approach and the prototype flame are discussed. Also the turbulence chemistry interaction modeling, the thermodynamic properties, the transport properties and artificial thickening are discussed. Then, a section is dedicated to emissions prediction, and the last section describes the fundamentals of the acoustic prediction.

4.2 The CFD Solver

The CFD solver provides the framework for the calculation of the different physical effects. The basis of the applied LES solver relies on a solver from the *OpenFOAM* [85] open source package. The detailed adjustments to the native large eddy simulation (LES) solver are explained in the following paragraphs. Thereafter, the next chapter describes the customizations for the combustion modeling.

To solve the compressible Navier Stokes equations, numerical methods are utilized. The Navier Stokes transport equations are of a non-linear partial differential type [86]. To obtain a solution for these transport equations and to keep the computational effort within reasonable limits, simplification, modeling and discretization methods were applied. Afterwards, a solution was obtained based on the initial and boundary conditions.

Use of simplification methods lead to the final type of equations and their solving algorithm. A typical simplification is the neglect of the temporal terms to get steady equations, but due to the transient character of the physical effects of interest, this simplification could not be applied. The whole set of compressible Navier Stokes equations had to be solved but the set of species equations was limited for simplification. To solve the transient Navier Stokes equations, two approaches were established; namely, direct numerical simulation (DNS) and LES. While DNS does not require turbulence modeling [87], LES acts as a low-pass filter, which makes modeling of small turbulent scales necessary. LES was chosen for this work because the computational effort is significantly lower than for DNS since DNS requires a very high spatial and temporal resolution. Furthermore, reacting DNS may require the modeling of all involved species which is a tremendous computational extra effort.

Most of the modeling assumptions that are made are described in the next sections. Nevertheless, the turbulence modeling is explained within next paragraphs because it is not only combustion related. For LES, the SGS viscous stress tensor ($\widetilde{\Pi}_{ij,sgs}$) out of the momentum equation, Eq. (2.18), has to be modeled. Eq. (4.1) shows the modeling approach according to the hypotheses of Boussinesq [88] for varied density flows. The second term on the right hand side ($\frac{1}{3}\delta_{ij}\Pi_{kk}$) represents a pseudo pressure term by using the Kronecker-Delta symbol (δ_{ij}); details on this are shown in the book of Poinsot and Veynante [11]. However, ν_{sgs} in the first term needs modeling effort. ν_{sgs} represents SGS viscosity term which may be modeled by an algebraic model or based on a transport equation.

$$\widetilde{\Pi}_{ij,sgs} \triangleq \widetilde{u_i u_j} - \widetilde{u}_i \widetilde{u}_j = -\nu_{sgs}\left(\frac{\partial \widetilde{u}_i}{\partial x_j} + \frac{\partial \widetilde{u}_j}{\partial x_i} - \frac{2}{3}\delta_{ij}\frac{\partial \widetilde{u}_k}{\partial x_k}\right) + \frac{1}{3}\delta_{ij}\Pi_{kk} \quad (4.1)$$

A well-established approach for the modeling of the SGS viscosity term ν_{sgs} is based on Smagorinsky [89]. The Smagorinsky approach relies on an algebraic equation. More sophisticated models use transport equations for the kinetic SGS energy (k_{sgs}) [90]. The advantage of transport equations is an imposed memory to the transported turbulence quantities [91]. The one equation eddy approach was used in this work in a non-dynamic version [90]. This model was selected for this work since relatively coarse meshes are used to enable efficient simulations of complex technical geometries and the one equation model is assumed to compensate for some discretization errors. Moreover, previous work (e.g. [92]) has demonstrated an advantage of the applied one equation eddy model in contrast to the Smagorinsky model for pressure loss prediction. The model parameters were defined according to the *OpenFOAM* default formulation [85]: $c_k = 0.094, c_e = 1.048$. Comparable values of $c_k = 0.07, c_e = 1.05$ were presented by Fureby et al. [93]. Fureby et al. [94] explained in a later publication the transport of the kinetic SGS energy Eq. (4.3)[1] according to the formulation of Schumann [95]. The kinetic SGS energy (k_{sgs}) is than used in Eq. (4.2) in order to calculate the SGS viscosity term.

$$\nu_{sgs} = \widetilde{k_{sgs}}^{1/2} c_k \Delta \quad (4.2)$$

---

[1]The local cell size is approached by the cube root of the cell volume according to the following formula: $\Delta = \sqrt[3]{V_{Cell}}$.

$$\frac{\partial \tilde{k}}{\partial t} + \frac{\partial \left(\tilde{u}_i \tilde{k} \right)}{\partial x_i}$$

$$= \nu_{sgs} \left(\frac{\partial \tilde{u}_i}{\partial x_j} + \frac{\partial \tilde{u}_j}{\partial x_i} \right)^2 + \frac{\partial}{\partial x_i} \left[(\nu_l + \nu_{sgs}) \frac{\partial \tilde{k}}{\partial x_i} \right] + c_e \frac{1}{\Delta} \tilde{k}^{3/2}$$

(4.3)

The numerical accuracy is strongly affected by the LES filter width, which is used in the turbulence modeling. LES filters are implicitly related to spatial discretization since e.g. a low-pass filter width can be determined by the cell size as in this work. The basic idea behind LES is to model the computationally expensive low scale turbulence effects and to resolve most of the large scale turbulence. Pope [96] recommends to resolve 80 % of the turbulent kinetic energy. As mentioned in the previous paragraph, a one equation turbulence model is used in this work to account for the less accurate temporal discretization in limited regions of a complex domain. Therefore, Pope's recommendation may not be applied in some regions. This is acceptable for the type of problems discussed in this work, as proven for full-scale validation, Chapter 7. A highly possible justification for the exception from Pope's rule is that the time inaccurate regions are very small in comparison to the whole domain. Moreover, those regions are not within the flame zone. The applied filtering operator $\Omega(\vec{x})$ is described in Eq. (4.4); x^* represents the location of the designated cell center.

$$\Omega(\vec{x} - x^*) = \begin{cases} \frac{1}{V_{Cell}} & \text{if } |\vec{x} - x^*| \leq 0.5\Delta \\ 0 & \text{else} \end{cases}$$

(4.4)

Due to consideration of variable density, Favre averaging is used in order to avoid unintended additional correlations. Eq. (4.5) shows how an arbitrary Favre averaged quantity $\tilde{\varphi}(\vec{x}, t)$ is obtained as a function of the LES filter.

$$\overline{\rho} \tilde{\varphi}(\vec{x}, t) = \int_V \rho \Omega(\vec{x} - x^*) \varphi(\vec{x}, t) \, dx$$

(4.5)

Spatial and temporal discretization is applied to the transport equations to get an approximated solution by using numerical methods. Based on initial and boundary conditions a solution is generated iteratively until the convergence is appropriate. Table 4.1 gives an overview of the most important terms of the transport equations and the applied discretization schemes. For temporal discretization the Crank-Nicolson method [97] was used in a modified version. The modification blends the original Crank-Nicolson method of purely 2nd order with an Euler method of 1st order. A blending factor of 0.5 was used to achieve an appropriate compromise between numerical stability and accuracy. While 1st order methods are known to support stability, 2nd order methods are used to increase accuracy (i.e. reduce the discretization error).

Table 4.1: Discretization schemes for different terms of the transport equations.

	Scheme	Order
Temporal	Crank-Nicolson	1st / 2nd Order blend
Convective (default)	Central	2nd Order
Convective (TKE[1])	Upwind	1st Order
Convective (scalars)	Upwind, face limited	2nd Order
Convective (velocity)	TVD[2] central	2nd Order
Diffusive	Central	1st / 2nd Order blend
Gradient (default)	Gauss	2nd Order
Gradient (velocity)	Gauss, face limited	2nd Order

As for the temporal discretization, the spatial and gradient discretization schemes were chosen as a compromise between accuracy and stability based on experience with GT-relevant cases. Especially the application of a face limiter on the convective term provides numerical stability with just minor effects on accuracy occurring. The limiter as well as the total variation diminishing (TVD) scheme allow sharp gradients to be resolved on coarse grids. Those gradients may occur in a transonic environment as is common in some regions of a GT. Only for the transport of turbulent kinetic energy (TKE) was a 1st order scheme used. The induced numerical error of the TKE is tolerated because this approximation provides still better predictions than those obtained by using the Smagorinsky approach (i.e. without a TKE transport equation). The solutions are obtained on the cell centers by using a finite volume method. Further details are described in the Ph.D. Thesis of Jasak [98].

---

[1] Turbulent kinetic energy (TKE)
[2] Total variation diminishing (TVD)

To finally resolve the problem of the numerical formulation of the differential equations, initial and boundary conditions have to be defined. Table 4.2 shows how the boundaries were chosen by default. 'Zero gradient' describes a Neumann type boundary condition, where the derivative (perpendicular towards the surface of the boundary condition) is zero. In contrast, 'fixed value' boundary conditions are of the Dirichlet type. In case of the velocity field, the 'fixed mass flow' boundary condition models the velocity field on the boundary according to the density. The density is obtained by the ideal gas law. For the viscosity calculation based on the TKE, wall functions were used to allow for grids without boundary layer refinement. Boundary layer refinement would increase the computational costs and reduce the robustness.

Table 4.2: Default boundary conditions for the LES approach.

	Inlet	Outlet	Wall
Velocity	fixed mass flow [1]	zero gradient	no slip
Pressure	zero gradient [2]	fixed value [3]	zero gradient
Temperature	fixed value	zero gradient	zero gradient [4]
Species	fixed value	zero gradient	zero gradient
Viscosity	-	-	wall function

In order to account for coupling between the pressure and velocity transport equations in the compressible flow environment, an appropriate solving algorithm has to be chosen. As mentioned before, one intention of this work was to provide a CFD based method for the prediction of combustion phenomena in complex geometries with reasonable computational costs, but complex geometries comprise a huge variety of length scales (e.g. fuel injection holes vs. flame ducts). To take the smallest scales into account without compromising the computational costs, the ability for local temporal compromises in accuracy should be provided. The numerical stability as a function of the grid size and the time step can be assessed by means of the CFL-number,

---

[1] With an optional artificial turbulence generator according to Klein [99]
[2] Acoustically fully-reflecting
[3] Acoustically nonreflecting
[4] Zero gradient is used in case of an adiabatic problem. In case of a diabatic problem, fixed values are used.

Eq. (2.24). Using an explicit formulation of the pressure influence on the momentum equations would prevent convergence if the CFL number were above one [5]. Therefore, an implicit or semi-implicit (i.e. without consideration of the neighboring velocities) formulation is of interest. The SIMPLE (semi-implicit method for pressure-linked equation) algorithm [86] was used to allow for CFL numbers above one. The algorithm consists of an iterative loop for the pressure correction within a time step. For the implicit solution of the transport equations, three internal iterations per time step were used. A relaxation factor of 0.6 for every transport quantity was used to increase the numerical stability. Using an explicit solver, just one internal iteration would be necessary but numerically stable calculations are limited to a convective CFL number of max. 0.3 for the designated technical cases. Consequently, the implicit solver setting results in a faster solution for convective CFL numbers above one. This is because, for the implicit approach, the same number of transport equations need to be solved at a CFL number of 0.9. The implicit solver would need one time step with three internal loops, while the explicit solver would need three time steps without internal loops at a CFL number of 0.3.

4.3 Combustion modeling approach

4.3.1 Prototype flames for tabulated chemistry

1D flames are used to generate a set of prototype solutions in advance. Subsequently, those solutions (in particular the source terms of the governing species) are made accessible, via look-up tables, for the transport equations to close the source term modeling. This approach is preferred in contrast to a flame speed closure approach because the source term has complex, i.e. three-dimensional behavior in the mixture fraction and reaction progress space. Fig 4.1 shows for example how the CO_2 source term may evolve in this space. A flame speed closure would not cover this three-dimensional effect because for a flame speed closure approach the origin of the source depends only on the change of the reaction progress.

According to the flamelet approach the 1D-FPLF was chosen as the prototype flame. As mentioned in the previous chapter, this type of flame is representative for a flamelet at GT-relevant conditions. In the tabulated chemistry

---

[5]In practical cases CFL numbers close to one are already numerically unstable.

context, detailed chemistry can be used as a basis for the prototype flames without influencing the computational efficiency of the CFD. As justified in the previous chapter, the GRI 3.0 mechanism [21] was utilized to cover the combustion of methane and natural gas in the present work. For the simulation of the prototype flame the software package *Cantera* [64] was applied. Three controlling variables were used to establish mapping of the prototype flames to the CFD: the reaction progress of the governing species c_{gov}, the mixture faction Z, and the heat loss quantity β. In the present work the reaction progress is calculated with the governing species CO_2.

For the governing CO_2 mass fraction a transport equation was solved based on the source term from the tables. For a prototype flame, one-dimensional transport equations of every species of the mechanism are solved. This has been explained in the fundamentals (Chapter 2) in more detail; Eq. (4.6) shows an example of a transport equation of a species mass fraction without the simplifications, made for the laminar prototype flame (Eq. (2.5)). The source term $\dot{\omega}_k$ in the transport equation is the chemical production rate created by an Arrhenius type equation. The prototype flames are also used to get the equilibrium of Y_{CO_2} as well as the thermal flame thickness[2] and the laminar flame speed as a function of the mixture fraction in advance. These values were also stored in the tables to support the thickening.

$$\rho \frac{\partial \mathbf{Y_k}}{\partial t} + \rho u \frac{\partial \mathbf{Y_k}}{\partial x_i} = \frac{\partial}{\partial x_i} \left(\rho D_k \frac{\partial \mathbf{Y_k}}{\partial x_i} \right) + \dot{\omega}_k \tag{4.6}$$

$$c_{gov} = \frac{Y_{CO_2}(Z, \vec{x})}{Y_{CO_2,eq}(Z)} \tag{4.7}$$

Y_{CO_2} was chosen as the only governing species to define the whole flame because CO_2 has a thicker flame brush than a linear combination of several species [62]. Therefore, a coarser mesh can be used to resolve the flame with a certain number of cells. Eq. (4.7) shows how the reaction progress of the governing species is determined by using its equilibrium value, which is a function of the mixture fraction. However, it should be mentioned that Y_{CO_2} is not strictly monotonic in rich regions ($\phi > 1.2$), which is problematic for the mapping of the source terms out of the look-up tables. Although this is beyond the scope of lean combustion modeling, it has to be considered for table generation. Therefore, $Y_{CO_2,eq}$ has been approximated by the maximum value

---

[2]$\delta_{th,l} = (T_{adia} - T_u)/max(dT/dz)$

of CO_2 for the rich cases, which do not exhibit a strictly monotonic behavior. Consequently, for some rich cases, the flame is modeled slightly more thinly than in reality. The limitations of the controlling variables are later shown in table 4.3 in combination with the post-flame emission modeling.

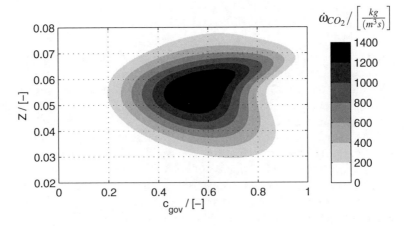

Figure 4.1: Governing CO_2 source term as a function of its reaction progress (Eq. (4.7)) and the mixture fraction (Eq. (2.1)) at $p = 3$ *bar*.

The applied method neglects strain and radiation. Strain becomes important for very lean or rich condition in a GT combustion system, using methane as fuel. The minor influence of strain for a lean flame at elevated pressure has been experimentally shown [100] for methane ($\phi = 0.67$), while for other fuel gas types the influence of strain was more significant. Radiation can become an important parameter [84] in dependence on the combustion system and fuel type and is neglected for simplification. Because of these assumptions, the application range of the model is limited to methane combustion.

The heat loss quantity β is, besides the progress and the mixture fraction, the third controlling variable of the tables. However, a separate investigation in Chapter 6 suggested no further consideration of the heat loss. This separate study showed very small influence of β on the source term at conditions which are relevant for a heavy-duty GT. This can be explained by elevated pressure conditions which make the energy density high in comparison to the heat loss in a moderately cooled system. Therefore, the functionality of the heat loss influence on the source term is included in the model description, but it was only applied in the separate study to justify the negligence. However, for atmospheric lab-scale systems the heat loss influence should be revised again, as demonstrated in the literature [101]. The heat loss influence on the source term has to be distinguished from the heat loss influence on the flow field, which is always considered within this work based on the energy equation. The energy equation reflects the transport of the total enthalpy. Consequently, heat loss over walls decreases the local temperature.

The heat loss quantity β, as the controlling variable, is defined in Eq. (4.8) as one minus the total enthalpy loss (Δh_{total}) related to the enthalpy of the heat release of a stoichiometric methane/air mixture. This stoichiometric heat release can be obtained by multiplying the lower heating value (LHV) by the stoichiometric mixture fraction of $Z_{st} = 0.055$ for methane.

$$\beta = 1 - \frac{\Delta h_{total}}{Z_{st}\text{LHV}_{CH_4}} \qquad (4.8)$$

The look-up tables were generated in a two-dimensional format for the reaction progress and the mixture fraction space. The third dimension (heat loss) was realized by usage of separate tables. A bi-linear interpolation method was used during the CFD calculations for the first two dimensions, while the closest value was chosen for the third dimension. The bi-linear interpolation ensures that even minor changes in the controlling variables are considered. The basic reaction progress space of the tables was resolved with 200 equidistant points and the mixture fraction space with an equidistant increment of 0.001. This accuracy of the tables is comparable to the accuracy identified in the literature [69]. Due to the minor role of heat loss for a secondary investigation, heat loss was resolved with five steps.

4.3.2 Transport of governing species

The CFD solver approximates the solution of the compressible Navier Stokes equations, the mixture fraction as well as the transport equation for the governing species Eq. (4.9). The latter was derived from Eq. (4.6). The bar operator denotes spatial filtering, while the tilde operator indicates a Favre averaged spatially filtered quantity. The right hand side of this equation has to be modeled.

$$\frac{\partial \bar{\rho}\tilde{Y}}{\partial t} + \frac{\partial \bar{\rho}\tilde{u}_i\tilde{Y}}{\partial x_i}$$

$$= \nabla \cdot \left(\overline{\rho D \nabla Y}\right) + \nabla \cdot \left(\bar{\rho}\widetilde{uY} - \bar{\rho}\tilde{u}\tilde{Y}\right) + \overline{\dot{\omega}_t} \tag{4.9}$$

The filtered laminar diffusion term $\nabla \cdot \left(\overline{\rho D \nabla Y}\right)$ was modeled by means of $\nabla \cdot (\bar{\rho} D \nabla \tilde{Y})$. It was shown [102] that neither filtering of the prototype flame nor more accurate modeling is necessary as long as the laminar flame thickness $1/max(|dc/dz|)$ is resolved within more than five cells. The same modeling approach was used in the energy equation. The $\nabla \cdot (\bar{\rho}\widetilde{uY} - \bar{\rho}\tilde{u}\tilde{Y})$ term represents the unresolved convection and was modeled through a gradient-based approach. The gradient approach $\nabla \cdot (\frac{\mu_{sgs}}{Sc_{sgs}} \nabla \tilde{Y})$ is well-established in the literature [103] and is consistent with momentum modeling. The SGS turbulent source term $\overline{\dot{\omega}_t}$ was modeled as a function of the laminar source term because the laminar source term is known from the prototype flames.

Based on a dimensional analysis (see Chapter 2), it can be shown that the laminar source term is proportional to the square of the laminar flame speed divided by the laminar thermal diffusivity (Eq. (4.10)). Assuming the same proportionality for a turbulent flame by using the effective flame speed and the effective thermal diffusivity ($a_{eff} = a_{SGS} + a_l$) leads to Eq. (4.11). In combination with Eq. (4.10), Eq. (4.12) can be derived. It has been shown before [68] that the laminar and turbulent source terms are proportional to their inverse time scales, which are proportional to the square of the associated laminar or turbulent flame speeds. The assumption for the modeling of the effective thermal diffusivity and the effective flame speed (S_{eff}) is that those quantities are laminar at zero SGS fluctuations because in this case the turbulence is resolved in time and space. The ratio between effective and laminar flame speed is proportional to the ratio of a real wrinkled flame surface to the modeled flame surface as shown [100] for the ratio between the laminar and turbulent flame speeds. This surface ratio is called a wrinkling factor Ξ. The derived equation shows the relationship between a laminar and a turbulent source term, with a potential increase of the source term due to wrinkling. However, inherent thickening of the flame due to SGS turbulence is also considered through the ratio of the thermal conductivities.

$$\dot{\omega}_l \propto \rho_u \frac{1}{\tau_l} = \rho_u \frac{S_l}{\delta_l} \propto \rho_u \frac{S_l^2}{a_l} \tag{4.10}$$

$$\dot{\omega}_t \propto \rho_u \frac{1}{\tau_t} = \rho_u \frac{S_{eff}}{\delta_t} \propto \rho_u \frac{S_{eff}^2}{a_{eff}} \tag{4.11}$$

$$\dot{\omega}_t = \frac{S_{eff}^2}{S_l^2} \frac{a_l}{a_{eff}} \dot{\omega}_l = \Xi^2 \frac{a_l}{a_{eff}} \dot{\omega}_l \tag{4.12}$$

Due to the non-linear behavior of the source term as a function of the re-action progress (i.e. space), the mean source term $\overline{\dot{\omega}_t}$ might be obtained by integration of the source term in SGS space for every time step. This would be necessary if the cell size was very coarse in comparison to the laminar flame thickness. However, due to the potential extra computational effort and the need to resolve the flame properly anyway, the source term was not integrated over the cells width as the integration interval. Instead, one can consider the source term proposed here as integrated with a Dirac delta function with a peak at the mean reaction progress of a cell. Proch [69] showed that this assumption is valid in the case of a thickened flame approach. It is important to consider SGS fluctuations separately from flame wrinkling, which results in a greater SGS flame surface. Wrinkling is discussed in a separate subsection.

Theoretically, there is also a dependency of the source term on the mixture fraction, and the heat loss quantity on the SGS level. However, these relations can be assumed to be linear as long as the variation is small and the flame is in the lean combustion regime. Assuming a Gaussian distributed PDF, the integration of such a linear dependency results in the mean value. Consequently, this dependency was not considered.

The above explained modeling approaches are the closed-form expression of Eq. (4.9) which is Eq. (4.13). It has to be considered that this modeling approach is justified as long as the flame is resolved properly, which may be ensured by the addition of artificial thickening. The addition of artificial thickening will be shown in a later subsection and is investigated in the verification section of Chapter 6. Later on, the turbulence chemistry interaction is investigated in a turbulent lab-scale validation case.

$$\frac{\partial \bar{\rho}\tilde{Y}}{\partial t} + \frac{\partial \bar{\rho}\tilde{u}_i\tilde{Y}}{\partial x_i}$$
$$= \frac{\partial}{\partial x_i}\left(\left(\bar{\rho}D + \frac{\mu_{sgs}}{Sc_{sgs}}\right)\frac{\partial \tilde{Y}}{\partial x_i}\right) + \Xi^2 \frac{a_l}{a_{eff}}\dot{\omega}_l \tag{4.13}$$

The mixture fraction transport, Eq. (4.14) was modeled using the same approximations as for the governing species. The mixture fraction is a conservative quantity because it describes the mass fraction of the elements of the fuel (i.e. the C and H atoms for methane combustion). The laminar diffusion coefficient D was approximately the same as for the governing species because the mixture faction was not related to a specific species and therefore the diffusion was not clearly defined. In the case of significant SGS turbulence, the laminar term will be small in comparison to the turbulent diffusion; this makes the potential error small. However, in the case of a laminar mixture fraction distribution or when approaching DNS with LES it might be important to investigate this further.

$$\frac{\partial \bar{\rho} \tilde{Z}}{\partial t} + \frac{\partial \bar{\rho} \tilde{u}_i \tilde{Z}}{\partial x_i} = \frac{\partial}{\partial x_i} \left(\left(\bar{\rho} D + \frac{\mu_{sgs}}{Sc_{sgs}} \right) \frac{\partial \tilde{Z}}{\partial x_i} \right) \tag{4.14}$$

4.3.3 Transport and thermodynamic properties

The transport and thermodynamic properties (TTP) have to be modeled for the CFD as for the laminar paradigm flame. The modeling of the turbulent stresses and fluxes have been explained in the subsections before but the corresponding laminar transport properties have to be closed as well. Namely, the laminar diffusion coefficient (D) as well as the laminar thermal diffusivity (a) and the laminar viscosity (μ) were modeled based on the mixture averaged approach [14–16] as used also for the prototype flame simulated by *Cantera* [64]. The mixture averaged approach is explained in the fundamentals (Chapter 2 for the laminar diffusion coefficient in Eq. (2.8).

In comparison to the prototype flame, the amount of species considered for the transport properties in the CFD were reduced. The species of an extended CH_4 one step reaction from Eq. (4.15) have been used to approximate the species of the whole mechanism for simplicity reasons. The advantage of this approach is that the species can be calculated based on algebraic equations and no further coupled transport equations have to be solved, which saves computational time and improves computational robustness. The species mass fractions are a function of the progress of the governing species and the mixture fraction. According to this approach, the diffusion coefficient e.g. represents the diffusion of one species into the remaining mixture. It should be pointed out that for the CFD based TTP only the mass fraction of CO_2 (Y_{CH_4}) was modeled by means of a transport equation while the other mass fractions of $CH_4, O_2, N_2, H_2O, H_2$ and CO were calculated. Those other mass fractions are just used for the calculation of the TTP and do not have to be mixed up with e.g. the emissions calculated by the transport equations. Therefore, those mass fractions are marked by the index TTP. The verification section in Chapter 6 shows that this approach is adequate for reproducing the laminar flame speed.

$$CH_4 + \frac{2}{\phi}\left(O_2 + \frac{0.79}{0.21}N_2\right)$$
$$\rightarrow \alpha H_2O + \beta H_2 + \gamma CO_2 + \delta CO + \frac{2}{\phi}\frac{0.79}{0.21}N_2$$

(4.15)

For the algebraic calculation of the mass fractions based on Eq. (4.15) the coefficients have to be approximated. The calculation of the coefficients $\alpha, \beta, \gamma, \delta$ is based on the water-gas shift reaction for rich conditions. However, for lean conditions, CO and H_2 do not significantly affect the thermo-dynamic and transport properties because their amount is small. For lean conditions H_2 and CO were assumed to be zero from the TTP point of view. As already mentioned, the intention was to provide a simplified analytical approach for the calculation of the main species of the extended one step CH_4 reaction. Nevertheless, the flow field CO emissions were also calculated based on a transport equation with a focus on the emissions without coupling to the analytical calculation of CO for the TTP as described here.

The controlling variables are the governing species CO_2 and the mixture fraction Z. Using Eq. (4.16) the remaining TTP species are calculated according to Eq. (4.17). The equilibrium constant K_p for the water-gas shift reaction is estimated to be 3. Actually, K_p varies with temperature but the slope is shallow. The resulting maximum deviation due to this assumption is about 10 % in the CO prediction, assuming a temperature range of $1600\,K \pm 400\,K$.

$$K_p = \frac{X_{CO}X_{H_2O}}{X_{CO_2}X_{H_2}} \approx 3$$
$$A = 2/\phi - (3(K_p - 2))/(2(K_p - 1))$$
$$B = 8(1 - 1/\phi)/(K_p - 1)$$
$$\alpha = 2 - \beta$$
$$\beta = max\left(0, \sqrt{A^2 + B} - A\right)$$
$$\gamma = 4\phi + \beta - 3$$
$$\delta = max(0, 4(1 - 1/\phi) - \beta)$$

(4.16)

$$Y_{CO,\text{TTP}} = Y_{CO_2} \frac{M_{CO}}{M_{CO_2}} \frac{\delta}{\gamma}$$

$$Y_{H_2,\text{TTP}} = Y_{CO_2} \frac{M_{H_2}}{M_{CO_2}} \frac{\beta}{\gamma}$$

$$Y_{CH_4,\text{TTP}} = Z - \left(\frac{Y_{H_2,\text{TTP}}}{M_{H_2}} + \frac{Y_{CO,\text{TTP}}}{M_{CO}} + \frac{Y_{CO_2}}{M_{CO_2}} \right) M_{CH_4}$$

$$Y_{H_2O,\text{TTP}} = 2\left(Z - Y_{CH_4,\text{TTP}}\right) \frac{M_{H_2O}}{M_{CH_4}} - Y_{H_2,\text{TTP}} \frac{M_{H_2}}{M_{CH_4}}$$

$$(4.17)$$

$$Y_{air} = 1 - Y_{CH_4,\text{TTP}} - Y_{H_2O,\text{TTP}} - Y_{H_2,\text{TTP}} - Y_{CO_2} - Y_{CO,\text{TTP}}$$

$$Y_{O_2,burnt} = \left(0.5 \frac{Y_{H_2O,\text{TTP}}}{M_{H_2O}} + 0.5 \frac{Y_{CO,TTP}}{M_{CO}} + \frac{Y_{CO_2}}{M_{CO_2}} \right) M_{CH_4}$$

$$Y_{N_2,\text{TTP}} = \frac{\left(Y_{air} + Y_{O_2,burnt} \right)}{1 + \frac{0.21}{0.79} \frac{M_{O_2}}{M_{N_2}}}$$

$$Y_{O_2,\text{TTP}} = Y_{air} - Y_{N_2,\text{TTP}}$$

The thermodynamic properties are also calculated based on the species of Eq. (4.15) as with the calculation of the transport properties. For those species, the so-called NASA coefficients [104] were stored to calculate the specific heat capacity. Consequently, the thermodynamic properties as well as the transport properties were modeled by the present approach as a function of the mixture and the temperature. This thermophysical approach enables the coupling of tabulated chemistry with the compressible CFD framework because of the temperature sensitivity of the properties.

4.3.4 Turbulence chemistry interaction

In this thesis, the turbulence chemistry interaction describes the influence of the SGS turbulence on the flame structure (i.e. wrinkling). Borghi [19] described different combustion regimes between wrinkled and thick flames for the whole macro-scale flame. However, in the case of LES, the flame gets filtered and not only the macro-scale is of interest but also the SGS. The flame structure on a SGS level can be evaluated for the determination of the SGS regimes [53]. Borghi's original regime description, as shown in Fig 2.2, can be used to determine a proper approach to model the ratio between effective and laminar flame speed (i.e. the SGS wrinkling).

To identify the regime of combustion on the SGS level, a turbulent SGS Damköhler number may be used. Eq. (2.16) can be used to derive Eq. (4.18) by using the SGS quantities. Indeed, the turbulent SGS length L_{SGS} can be determined by the cell size resp. the LES filter width Δ. Using the turbulent SGS velocities u'_{sgs} it follows the turbulent time scale τ_t. The laminar resp. chemical time scale τ_l is calculated by the ratio of laminar flame thickness $\delta_{th,l}$ over the laminar flame speed S_l. It can be assumed that the flame is resolved on a certain number of cells (e.g. ensured by artificial thickening of the flame). Therefore, the ratio between L_{SGS} and $\delta_{th,l}$ can be assumed to be below one. Assuming that the laminar flame speed is lower than the SGS velocity fluctuations u'_{sgs} it follows a turbulent Damköhler number smaller than one on the SGS level. Also this assumption is valid for typical LES simulations of GT-combustors because the laminar flame speed of methane is typically below 1 m/s, while the velocity fluctuations are typically above (see for example Fig. 6.6).

$$Da_{t,sgs} = \frac{\tau_t}{\tau_l} = \frac{L_{sgs}/u'_{sgs}}{\delta_{th,l}/S_l} = \frac{L_{sgs} \cdot S_l}{\delta_{th,l} \cdot u'_{sgs}} \tag{4.18}$$

According to Borghi, Damköhler called the resulting regime – below a Damköhler number of one – "distributed combustion", which can be considered for use with SGS modeling even if the macroscopic scales are in another regime [53]. Consequently, the distributed combustion regime was considered to describe the SGS wrinkling factor Ξ.

Schmid et al. [68] tried to cover the main combustion regimes with Eq. (4.19) for RANS simulations as a function of the macroscopic turbulent Damköhler number. For small turbulent Damköhler numbers, Eq. (4.20) can be derived based on the turbulent Reynolds number Re_t [68]. Before, this model had been originally formulated by Damköhler [105] in 1940.

$$\frac{S_t}{S_l} = 1 + \frac{u'}{S_l} \left(1 + Da_t^{-2}\right)^{-1/4} \tag{4.19}$$

$$\frac{S_t}{S_l} \propto 1 + \sqrt{Re_t} = 1 + \sqrt{\frac{u'L_t}{a_l}} \Bigg|_{Da \ll 1} \tag{4.20}$$

41

Other authors described comparable formulations by considering the Lewis number based on experiments [106, 107] to cover a wider range of fuel gases. Therefore, Eq. (4.21) was used in the present work to close the turbulence chemistry interaction. In contrast to Eq. (4.20), Eq. (4.21) is formulated for LES and relies on the SGS fluctuations u'_{SGS} instead of the macroscopic turbulent scales u'. It is claimed that the model parameter C_Ξ of the wrinkling factor is valid for a broad range of grid sizes and operating conditions. This factor is determined in the validation section of Chapter 6. The index eff indicates the effective flame speed, consisting of the laminar plus the SGS part.

$$\Xi = \frac{S_{eff}}{S_l} = 1 + \frac{C_\Xi}{Le} \sqrt{\frac{u'_{SGS} L_{SGS}}{a_l}} \tag{4.21}$$

An accurate prediction of the flame position (i.e. turbulent flame speed) is important for predicting thermoacoustic phenomena. These phenomena can be caused by the time lag between the heat release center and fuel injection [43] and therefore on the flame position. For consolidation, all components of the selected modeling approach are successively investigated regarding their impact in the validation Chapter 6.

4.3.5 Introduction of a novel thickening approach

Artificial flame thickening was introduced [63] to achieve a properly resolved flame, and has been modified over recent years. The idea is to thicken a flame brush without influencing the flame speed in order to limit the size of the computational mesh. After the ATF was introduced to a LES environment [11], dynamic flame detection was utilized [108]. Dynamic flame detection limits the artificial treatment of the corresponding transport equation to the flame region. Later, grid adaptive thickening was applied [62] in order to adjust the thickening parameters as a function of the mesh.

The models described in the literature do not distinguish between inherent SGS thickening and explicit artificial thickening, which will be addressed in this section. However, thickening of the laminar flames itself has been shown before [102] by filtering of the laminar prototype flames using a LES filtering approach. In contrast, the present work introduces thickening of the laminar prototype flames using the original approach [63] as an input to the turbulent transport of the governing species. This approach is less complex than applying a filter to the laminar flames.

To extend the species equation for further artificial thickening without influencing the unresolved convection term, only the laminar part of the transport equation was thickened in the present work. Hence, a new laminar thickening factor F_l has been introduced. In Eq. (4.13) the original turbulent transport of the governing species without applying the ATF approach is shown. This equation distinguishes between the laminar diffusion and the unresolved convection term. Furthermore, the source term is modeled based on the laminar source term. Due to the explicit contribution of the laminar terms, the laminar terms can be modified by the laminar thickening factor without influencing the unresolved convection term as will be shown in Eq. (4.23). This approach requires accurate modeling of the laminar diffusion term because the interaction between the source term and the diffusion determines the flame speed [63]. This was addressed for the presented CFD solver.

The explained extension of the known ATF approach aims at explicit thickening of the laminar prototype flame without influencing the SGS models. To get an understanding of the novelty, the non-thickened species transport Eq. (4.13) is compared to a recent formulation from the literature [109] including thickening (Eq. (4.22)). The variable E represents the efficiency function, which accounts for turbulence chemistry interaction and F is the thickening factor.

$$\frac{\partial \bar{\rho}\tilde{Y}}{\partial t} + \frac{\partial \bar{\rho}\tilde{u}_i\tilde{Y}}{\partial x_i} = \frac{\partial}{\partial x_i}\left(\bar{\rho}F \, E \, D\frac{\partial \tilde{Y}}{\partial x_l}\right) + \frac{E}{F}\dot{\omega}_l \qquad (4.22)$$

Comparing the source term of Eq. (4.22) with Eq. (4.13) leads to an efficiency of $E = \Xi^2$, in accordance to the original thickening approach. Furthermore, an inherent thickening factor could be $F = a_{eff}/a_l$. This inherent turbulent thickening due to SGS fluctuations corresponds to the unresolved convection term. However, in comparison with the common thickening approach, this term is not explicitly considered in the diffusion term of Eq. (4.13). In consequence it also accounts for the turbulence chemistry interaction. Moreover, the efficiency function or the wrinkling factor, proposed in this work, influence only the source term. However, the efficiency function of Eq. (4.22) also influences the unresolved convection term. In contrast, as explained before, the well-established [103] gradient approach [3] was chosen in the present

---
[3] $\nabla \cdot \left(\frac{\mu_{sgs}}{Sc_{sgs}}\nabla\tilde{Y}\right)$

work for the modeling of the unresolved convection term. The advantage is its consistency with momentum transport modeling.

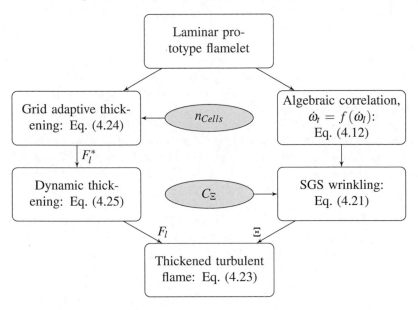

Figure 4.2: Novel thickening procedure with model parameters in gray clouds.

Fig. 4.2 shows how the novel thickening approach is utilized for the thickening of the laminar prototype flame and how it is aligned to the flame wrinkling and the tabulated chemistry. The advantage of the presented formulation is that inherent turbulent SGS thickening will be considered without an extra modeling effort and without artificial terms. As presented in the flow chart, the proposed combustion modeling approach relies on two model parameters, which will be determined in the validation section.

$$\frac{\partial \bar{\rho} \tilde{Y}}{\partial t} + \frac{\partial \bar{\rho} \tilde{u}_i \tilde{Y}}{\partial x_i}$$
$$= \frac{\partial}{\partial x_i} \left(\left(\bar{\rho} F_l D + \frac{\mu_{sgs}}{Sc_{sgs}} \right) \frac{\partial \tilde{Y}}{\partial x_i} \right) + \Xi^2 \frac{a_l}{a_{eff}} \frac{\dot{\omega}_l}{F_l} \tag{4.23}$$

According the flow chart in Fig. 4.2 the laminar thickening factor F_l is determined as a function of the grid. Thus, the grid adaptive thickening approach of Kuenne et al. [62] was used in Eq. (4.24). The maximum thickening term F_l^* was determined based on the choice of a certain number of cells n_{Cells} to resolve the already inherently thickened flame. Therefore, a comparison between the thermal laminar flame thickness ($\delta_{th,l} = (T_{adia} - T_u)/max(dT/dz)$) and cell size was carried out for every cell and time step to check if the flame would be resolved on the chosen number of cells n_{Cells} or if the flame needed further artificial thickening.

For the estimation of the local flame thickness in the CFD calculation, the thermal laminar flame thickness was tabulated as a function of the mixture fraction. Then, this thickness was multiplied by the inherent SGS thickening factor to get the turbulent thickness. A proper choice of the minimum number of cells to resolve the flame is described in the verification section of Chapter 6. Eq. (4.24) is valid for laminar cases as well as for turbulent cases. For laminar calculations, the effective thermal diffusivity is equal to the laminar diffusivity; consequently, only laminar thickening is considered. Indeed, if the inherent turbulent SGS thickening for turbulent flames is already thick enough, no further laminar thickening will be used. L_{SGS} is the characteristic SGS length scale calculated as the cube root of the corresponding mesh cell volume.

$$F_l^* = max\left(1, n_{Cells}\frac{L_{SGS}}{\delta_t}\right) = max\left(1, n_{Cells}\frac{L_{SGS}}{\delta_{th,l}\frac{a_{eff}}{a_l}}\right) \qquad (4.24)$$

Moreover, dynamic thickening was applied to make sure that artificial thickening was used only in the flame region. Therefore, a flame sensor was used according to the definition of Durand and Polifke [108] as shown in Eq. (4.25). The flame sensor indicates the change of the reaction progress c_{gov} of the governing species.

$$F_l = 1 + (F_l^* - 1)16\left[c_{gov}(1 - c_{gov})\right]^2 \qquad (4.25)$$

In the present subsection a new and straightforward procedure for thickening was derived which distinguishes between laminar and inherent turbulent SGS thickening. In comparison to previous work, inherent SGS thickening was not artificially created. Only the laminar prototype flame was thickened before serving as an input to the turbulent transport equation.

4.4 Emissions prediction

4.4.1 Modeling approach

To describe the emissions the relevant physical effects have to be identified and modeled. The objective of this thesis was to model CO and NO_x emissions. As already mentioned in Chapter 3, solving a detailed mechanism in a transient CFD code, covering CO and NO_x chemistry, would be prohibitively expensive for technical systems. However, fast chemistry effects introduced by radicals are crucial for the prediction [35, 36] of CO pollutants. Therefore, tabulated chemistry was chosen as a compromise and to be consistent with the modeling of the governing species. To account for the different time scales between the flame front and post-flame zone, both zones were modeled separately, as proposed for the time scale separation (TSS) approach [70]. Since previous work [5] showed the importance of transient effects on the mixture fraction distribution, the emissions' source terms were modeled as a function of the post-flame progress and the mixture fraction. Large scale fluctuations were resolved by the LES approach, while the SGS influence on the source terms was neglected for the same reasons as for the governing species (see previous section). For simplicity, heat loss and radiation influences on the source were not considered because they had minor effects on the relevant conditions (see Chapter 6). Furthermore, NO_x was approached by just NO itself because NO_2 is more than one order of magnitude smaller for GT-relevant conditions.

A flame can be divided into two zones: the flame front and the post-flame zone. Wegner et al. [70] introduced a method to distinguish between CO production and oxidation, which is also the basic approach used in this thesis. The present work extends the TSS method for transient calculations and NO prediction. Furthermore, the embedding in the combustion model itself was improved by using tabulated chemistry for the governing species of the whole flame instead of a flame speed closure approach.

A proper choice of the split between the two zones is crucial for the modeling of the separation approach. Due to the non-monotonic behavior of the CO mass fraction over the flame, the CO source term can be distinguished between net CO production and oxidation. Fig. 4.3 gives an example for the resulting two zones at very lean conditions to illustrate the relatively thick post-flame. Defining the CO oxidation path as the post-flame zone has two advantages: it enables a post-flame reaction progress variable to be defined based on the monotonic net CO oxidation, and it considers the different characteristic time scales. The reaction progress variable has to be monotonic to enable unique addressing of the source term. The peak in Fig. 4.3 of the CO mass fraction is slightly upstream of the position where the source term is zero. This can be explained by diffusion within the laminar prototype flame, which causes the transport of CO mass towards the inlet.

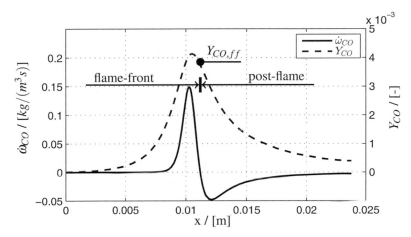

Figure 4.3: Exemplary prototype flame result: CO mass fraction and source term as a function of spatial coordinate for $\phi = 0.14$, $p = 20\ bar$ and $713\ K$ fresh gas temperature.

As stated in the previous paragraph, Eq. (4.26) defines the post-flame progress between the flame front value of Y_{CO} where the source term becomes zero and the equilibrium of CO. This post-flame progress was used as an index for both the CO and the NO transport. However, in contrast to the negative CO source term in the post-flame, the NO source term is positive and constant after a certain degree of progress, while the CO emissions are assumed to reach equilibrium values for most cases. In consequence, the CO source term will be zero at 100 % post-flame progress, while the NO source term remains constant. This approach is valid for typical lean partially premixed GT conditions with NO emissions below 100 $ppmv$ because the equilibrium emissions are an order of magnitude higher.

$$c_{pf} = \frac{Y_{CO}\vert_{\dot{\omega}_{CO}=0}(Z) - Y_{CO}(Z,\vec{x})}{Y_{CO}\vert_{\dot{\omega}_{CO}=0}(Z) - Y_{CO,eq}(Z)} \qquad (4.26)$$

The flame front progress was calculated based on the progress of the governing species (i.e. the whole flame) and based on the certain position where the source term of CO becomes zero. Resolving the whole flame rather than resolving the flame front separately helps to reduce numerical uncertainties since the whole flame is thicker. Hence, c_{ff} is calculated based on the progress of the governing species, as follows.

$$c_{ff} = \frac{c_{gov}(Z,\vec{x})}{c_{gov}\vert_{\dot{\omega}_{CO}=0}(Z)} \qquad (4.27)$$

4.4.2 Implementation

For the implementation of the previously presented modeling approach into the CFD package *openFOAM* [85], the transport equations for CO and NO were defined while considering the flame front and the post-flame source term. The purpose of the flame front source term is to reach a certain flame front mass fraction within the flame front thickness. The flame front mass fractions' were defined based on the corresponding mass fractions' at the location were the CO source term reaches zero as already illustrated in Fig. 4.3. This happens in a few cells for practical cases because the flame is assumed to be resolved on the number of cells defined in the thickening approach. Due to the thin flame front zone, a robust turbulent flame speed closure approach [49] was used for the CO and NO flame front source term closure as in the original TSS approach [70]. However, the post-flame zone was resolved by using tabulated chemistry, as already introduced for the governing species of the whole flame.

Eq. (4.28) shows the transport equation of CO, which differs from the governing species CO_2 (Eq. (4.23)) by the source terms and the absence of thickening. Thickening is not relevant for the CO and NO flame front progress because the pathway until the flame front emissions value is considered as not relevant for the integral emissions at the outlet of a technical combustion system. However, it is important to provide the flame front CO and NO values at the correct position as a starting point for the slowly reacting post-flame zone. For the post-flame, the CO oxidation is assumed to be slow enough to be resolved without artificial thickening.

The post-flame look-up method for the CO source term is implemented similarly to the look-up of the CO_2 source term, because for both cases the controlling variables are the mixture fraction and the reaction progress. The same ideas are applied for the transport of NO, which is described in Eq. (4.29).

$$\frac{\partial \bar{\rho}\widetilde{Y_{CO}}}{\partial t} + \frac{\partial}{\partial x_i}\left(\bar{\rho}\tilde{u}_i\widetilde{Y_{CO}}\right) - \frac{\partial}{\partial x_i}\left(\left(\bar{\rho}D + \frac{\mu_{sgs}}{Sc_{sgs}}\right)\frac{\partial \widetilde{Y_{CO}}}{\partial x_i}\right)$$
$$= \dot{\omega}_{CO,ff}\big|_{c_{pf}=0} + \dot{\omega}_{CO,pf}\big|_{c_{ff}=1} \tag{4.28}$$

$$\frac{\partial \bar{\rho}\widetilde{Y_{NO}}}{\partial t} + \frac{\partial}{\partial x_i}\left(\bar{\rho}\tilde{u}_i\widetilde{Y_{NO}}\right) - \frac{\partial}{\partial x_i}\left(\left(\bar{\rho}D + \frac{\mu_{sgs}}{Sc_{sgs}}\right)\frac{\partial \widetilde{Y_{NO}}}{\partial x_i}\right)$$
$$= \dot{\omega}_{NO,ff}\big|_{c_{pf}=0} + \dot{\omega}_{NO,pf}\big|_{c_{ff}=1} \tag{4.29}$$

As mentioned before, the pollutant flame front source terms are aimed at the preparation of a certain flame front mass fractions within the flame front thickness. Therefore, a robust flame speed closure approach was utilized to provide a universal flame front source term, which can be used for the NO and CO prediction. The flame speed closure approach, using the gradient of the reaction progress variable as a source, has been discussed in detail by Zimont [49]. The concept behind the flame speed closure is to provide a source term which ensures a certain turbulent flame speed of the reaction. In the present case the turbulent flame speed is approached by the laminar flame speed times the wrinkling factor (S_l times Ξ).

Eq. (4.30) shows the applied approach as a function of the laminar flame speed, which was obtained from the look-up tables. The integral over the gradient of the flame front progress ∇c_{ff} is one because c_{ff} changes from zero to one[4]; the integral and the gradient cancel each other out. Consequently, a certain flame front mass fraction is reached after 100 % flame front progress, if the universal, turbulent, flame front source term $\dot{\omega}_{t,ff}$ is multiplied by the desired flame front mass fraction.

---

[4] $\int_0^\infty \nabla c_{ff}dx \hat{=} 1$

To improve numerical stability, source term linearization was used to model the universal, turbulent, flame front source term $\dot{\omega}_{t,ff}$. Eq. (4.30) shows how the product rule was utilized for this purpose.

$$\dot{\omega}_{t,ff} = \rho_u \Xi S_l \nabla \left(c_{ff} \right) = \nabla \left(\rho_u \Xi S_l c_{ff} \right) - \nabla \left(\rho_u \Xi S_l \right) c_{ff} \qquad (4.30)$$

In Eq. (4.31) and Eq. (4.32) the specific pollutant flame front source terms are formulated according to the previously mentioned approach to ensure the desired flame front emissions after 100 % flame front progress. As described in the previous section, the flame front fractions were obtained from the look-up tables as a function of the mixture fraction. The certain flame front mass fractions of both pollutants were defined as the emissions at the position where the net CO oxidation starts; i.e. where the CO source term changes from positive to negative.

$$\dot{\omega}_{CO,ff} = \dot{\omega}_{t,ff} Y_{CO,ff} \qquad (4.31)$$

$$\dot{\omega}_{NO,ff} = \dot{\omega}_{t,ff} Y_{NO,ff} \qquad (4.32)$$

The limit values of the post-flame source terms as a function of the controlling variables are shown in table 4.3 in comparison to the CO_2 modeling. As discussed in the previous subsection, the CO source term becomes zero while the NO term becomes constant in the post-flame. Furthermore, the same lean and rich extinction limits were used as for the governing species CO_2. The lean limit was chosen to be $\phi_{lean} = 0.2$ in order to get converged solutions of the prototype flames. The rich limit of $\phi_{rich} = 2$ was chosen because higher values are not of interest for lean, partially premixed calculations. The heat loss quantity β was utilized for the governing species with an extinction limit determined by the solution of the prototype flames. The laminar flame speed decreases with heat loss at the beginning of the flame, which results in flame extinction in very cold conditions at about $\beta_{extinction} \approx 0.75$.

4.5 Evaluation of thermoacoustic prediction method

The thermoacoustic behaviors and the numerical setup interact in a computational simulation because significant numerical damping of acoustic ampli-

Table 4.3: Limitations of tabulation

	$\phi < \phi_{lean}$, $\phi > \phi_{rich}$	$\beta < \beta_{extinction}$	$c = 0$, $c = 1$	$c_{pf} = 0$	$c_{pf} = 1$
$\dot{\omega}_{CO_2}$	0	0	0	—	—
$\dot{\omega}_{CO,pf}$	0	—	—	$\dot{\omega}_{CO}$	0
$\dot{\omega}_{NO,pf}$	0	—	—	$\dot{\omega}_{NO}$	$\dot{\omega}_{NO} = const.$

tudes may be added. Thus, the numerical methods have to be chosen with care. Indeed, a remaining interaction has to be assessed in order to exclude a significant influence of the numerics on the solution. The assessment of the chosen setup in terms of numerically induced error is shown in this section.

The outlet pressure is chosen to be non-reflecting by default. This is a reasonable assumption if the boundary is aligned downstream of a chocked nozzle, because such a nozzle acts as a non-reflecting boundary condition anyway. For other problems, the type of reflection has to be chosen with care. The fully reflecting inlet condition is chosen to achieve computational robustness and can be justified if the inlet surface is very small compared to the surrounding surface.

When simulating acoustic phenomena, it is important to consider the acoustic CFL number, which is defined in Eq. (4.33). The acoustic CFL number includes the convective velocity and the speed of sound to account for acoustic waves propagating through a domain.

$$CFL_{acoustic} = \frac{(|\mathbf{u}| + c)\Delta t}{\Delta x} \tag{4.33}$$

As the implicit time discretization allows for convective CFL numbers above one, the acoustic CFL number might be even higher. For a self-excited simulation, it must be ensured that the effect of the numerical error is negligible. Hence, the numerical error has to be significantly below the physical damping. Beside a reduction in amplitude, discretization can result in a phase error, which is called dispersion. It was shown before [42] that numerical dispersion becomes relevant at acoustic CFL numbers above two.

$$\textbf{Numerical error} = \frac{\max(u'|_{[0:\lambda]}) - \max(u'|_{[\lambda:2\lambda]})}{\max(u'|_{[0:\lambda]})} \qquad (4.34)$$

For the selection of the time step, the acoustic CFL number and its influence has to be estimated. Furthermore, the number of cells per wave length has to be considered. Therefore, the cell size in the flame region was chosen as representative of the following short evaluation of the whole domain. The numerical error is defined as the relative decrease in the velocity amplitude of an acoustic wave (u') per wave length λ in Eq. (4.34). This error is shown in Fig. 4.4 to evaluate the numerical setup which was described at the beginning of this chapter. For this evaluation, acoustic waves were simulated in a 1D domain (2x2x2000 cells) by superimposing acoustic waves to a convective mean velocity of 10 m/s. This setup is comparable to a previous investigation [42]. The wave length and acoustic CFL number were varied to show the influence of the acoustic CFL number and the wave resolution on the numerical error. Assuming a tolerable numerical error of 0.01, a maximal acoustic CFL number of two can be accepted at a resolution of about 1000 cells per wavelength. For the wavelength, the target frequency has to be considered. These thoughts are considered in the full-scale validation described in Chapter 7.

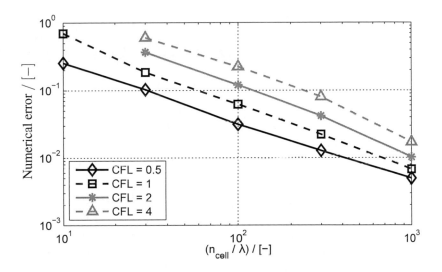

Figure 4.4: Numerical error as a function of the resolution and different acoustic CFL numbers. The investigation is valid for the numerical schemes, shown in Table 4.1 and used in the present work.

Chapter 5

Low order approach for NO_x emission prediction

5.1 Introduction

In contrast to the high-fidelity CFD-based approach the low order approach, described here, enables an extensive set of cases to be run. This is needed to perform optimizations or to create a correlation between mixing and emissions for a GT combustion system. Therefore, a generalized NO_x emissions model suitable for GT combustion was formulated according to a previous publication of the author [5], with minor changes. However, similar to the high fidelity approach the low order approach is based on a set of 1D-FPLF. The main difference is the domain: While the high fidelity approach is based on an extensive 3D mesh, the low order approach relies on an 1D approximation of a technical premixed GT combustion system.

The low order approach enables detailed chemistry modeling with a reasonable amount of computational effort based on the 1D-FPLF reactor. To reduce the computational effort further, tabulated chemistry was applied. However, to account for the reaction progress, a two-zone approach (flame front and post-flame zone) as sketched in Fig. 5.1, was considered. This is in accordance with the CFD-based approach, presented in the previous chapter. The equivalence ratio distribution as a result of the premixing was considered by means of a PDF.

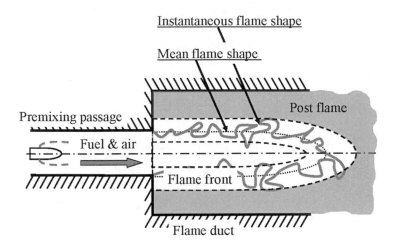

Figure 5.1: Schematic of flame zones in a technical combustion system

The NO_x source term was tabulated as a function of residence time and equivalence ratio in a look-up table. The contribution of the reaction progress in the flame front was taken into account via integration of the NO_x source term over the flame front residence time. In contrast, the post-flame source term of the flamelet was considered as constant. In contrast to the previously published model [5] the reaction progress as a table dimension was replaced by the residence time to be consistent with the consideration of the downstream mixing process as described in detail in the next section.

The downstream mixing was modeled based on a stochastic process where the standard deviation of the probability density function was modeled as a function of time. Even if the low residence time in the flame front permits the assumption of a constant probability distribution [5], with both zones (flame front and post-flame) it is considered that downstream mixing increases the validity. The cooling mass flows were considered through adding them to the flue gas but the influence on the source term was neglected.

5.2 Modeling approach

The modeling approach is described in three steps. First, the chemical kinetics and the flamelet model configuration are explained. Then, details of the integral 1D NO_x emission modeling, using the kinetic model, are described. Finally, a statistical description of a mixture distribution and evolution is given.

5.2.1 Chemical kinetics and the flamelet model

The objective of the modelling approach is to obtain NO_x formation source terms as a function of the residence time and the equivalence ratio. As described in Chapter 3, a flamelet approach allows a turbulent flame to be decomposed into separate laminar 1D flames. For a desired equivalence ratio range, a finite number of 1D flame calculations as a function of the intrinsic residence time can be calculated and tabulated. As for the CFD-based approach, the open source tool Cantera was used with the GRI-Mechanism 3.0 [21], which was justified by previous validation work for NOx prediction with methane fuel [72, 110]. Unlike the CFD-based approach, the low order model is able to consider both the NO and NO_2 part of the NOx emissions without compromising simplicity.

The chemical reactor model is the same as for the CFD-based approach: a 1D, laminar, perfectly premixed, freely propagating, flat flame. The applied method neglects strain, heat loss and radiation as with the CFD-based approach, and for the same reasons. For the combustion systems, considered in the present work, turbulence effects can be decoupled from NO_x kinetics for two reasons: the stiff flame front is thin in comparison to the turbulent length scales, and the NO_x source term is assumed to be constant in the post-flame region.

The resulting NO_x source terms of two exemplary 1D calculations are shown in Fig. 5.2 as well as the related temperature profile. The source term peaks are typical for laminar 1D-premix flames. In front of the main peak, there is a positive maximum and a negative minimum, which balance each other partially when integrated over the flame spread. At the end of the pathway, it can be seen that there is a constant source term. Hence, approximating space and time independence is appropriate in the post-flame region.

Based on the evolving NO_x source term, a laminar flame front thickness can be defined to distinguish between the flame front and the post-flame. The

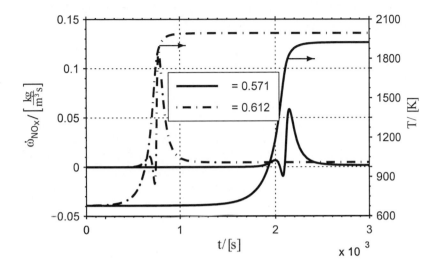

Figure 5.2: Exemplary temperature distribution and NO_x source term as a function of the laminar flame brush for two stoichiometric conditions (CH_4 at 8 *bar*)

reason for doing this is to identify the positions where the source term starts and stops changing. In Eq. (5.1) the thickness is defined as the distance between the last and first position of a distinctively changing source term over time. The change of the source term is defined by its derivation in time. For this work, the distinctively changing source term region is defined as the region where the source term changes at least by 1 % of its maximum change. This ensures that the source term is constant outside this range; i.e. in the post-flame region.

Fig. 5.2 shows an example of the source term propagation in the flame front for lean CH_4 reactions. Furthermore, the figure illustrates the dependency of the temperature gradient as a function of the equivalence ratio. The variable temperature gradient results in an equivalence ratio dependence of the flame thickness. The flame front definition of this section differs from the high order approach of Chapter 4 due to its functionality. The CFD-based flame front description relies on the progress of the CO formation, while here the region of a changing source term is considered.

$$\delta_{l,1D} = \Delta x = max \left(x|_{\frac{\dot{\omega}_{NOx}}{max(\dot{\omega}_{NOx})}=0.01} \right) - min \left(x|_{\frac{\dot{\omega}_{NOx}}{max(\dot{\omega}_{NOx})}=0.01} \right) \quad (5.1)$$

The table format for every calculated equivalence ratio is the NOx source term as a function of the residence time. For a preselected fuel and air temperature, as well as the desired pressure, 45 calculations have been performed for equidistant equivalence ratios between 0.2 and 2.0. Out of the range of the provided tables, the formation of NO_x emissions was set to zero. This is a reasonable approach because these regions are too lean for NO_x formation or too rich for relevance in a lean combustion system. Based on the laminar source terms and the resulting flame thickness, a NO_x model of a GT combustion chamber will be formulated in the next subsection.

5.2.2 Model description

As described in the introduction of this chapter, a global model for the NO_x formation has to be derived with the following input: Tabulated NO_x source terms, PDF of the flame front mixture, and residence time. The source terms in the table were parametrized by equivalence ratio and residence time. The conditional table, relying on discrete values, was then linearly interpolated.

Based on the modeled domain (divided into the flame front and the post-flame zone), the separate NO_x formation rates were calculated. A separate treatment was performed since the NO_x formation kinetics as well as the mixture homogeneity change with residence time. The modeling was simplified based on an assumption made in the previous subsection. There, the NO_x source term was described as constant over the post-flame progress.

The NO_x emissions at the end of the combustor were computed from the integral NO_x formation rate divided by the total mass flow including the cooling air (Eq. (5.2)). This approach does not account for the mixing of cooling air into the flame front or post-flame in terms of a changing equivalence ratio distribution, but for a residence time reduction due to the cooling air.

$$Y_{NO_x} = Y_{NO_x,FF} + Y_{NO_x,PF} = \frac{\dot{m}_{NO_x,FF} + \dot{m}_{NO_x,PF}}{\dot{m}_{cooling} + \dot{m}_{products}} \quad (5.2)$$

The flame front NO_x formation rate was calculated based on the flame front volume multiplied by the mean flame front source term (Eq. (5.3)). The mean flame front NO_x source term describes the formation of NO_x mass per volume and time. The volume can be described by the flame front surface multiplied by the mean laminar flame thickness $\delta_{l,1D}$ (Eq. 5.1). The NO_x source term, as a function of the equivalence ratio and the position, was derived from the tables. The inner integration was performed over the equivalence ratio to account for unmixedness. Outer integration over the time was performed until the end of the flame front was reached. This step considers downstream mixing in a stochastic process and is described in more detail for the post-flame zone.

Under the assumption of a known PDF distribution, the mean source term was determined based on the tabulated source distribution. An example source term distribution as well as a PDF distribution is shown in Fig. 5.3 for elevated pressure conditions. Eq. (5.3) shows how those fields are integrated over the equivalence ratio and the time. Furthermore, the flame front residence time obtained by Eq. 5.4 is shown in the right figure. Two points are remarkable: it is confirmed that the source term is constant on the right side of the curve (i.e. in the post-flame region), and the flame front residence time (i.e. the thickness) exponentially increases with leaner conditions. Therefore, downstream mixing is considered even for the flame front because it might be relevant for very lean conditions where a thick flame front might give enough time for an unmixedness reduction. The left plot of Fig. 5.3 shows a white line from the top left to the bottom right corner. In this area the source term is zero or slightly negative.

Multiplying the mean source term with the flame front volume (V_{FF}) in Eq. (5.3) leads finally to an emissions mass flow. The analytical calculation of the flame front volume as a function of the flame front thickness is described later. The flame front volume is not based on CFD.

$$
\begin{aligned}
\dot{m}_{NO_x,FF} &= V_{FF}\,\overline{\dot{\omega}_{NO_x,FF}} \\
&= \frac{V_{FF}}{t_{FF}} \int_0^{t_{FF}} \int_\phi \mathrm{PDF}\,(\phi,t)\,\dot{\omega}_{NO_x}(\phi,t)\,\mathrm{d}\phi\,\mathrm{d}t
\end{aligned}
\tag{5.3}
$$

$$
t_{FF} = \Delta t = max\left(t\Big|_{\frac{\dot{\omega}_{NO_x}}{max\left(\dot{\omega}_{NO_x}\right)}=0.01} \right) - min\left(t\Big|_{\frac{\dot{\omega}_{NO_x}}{max\left(\dot{\omega}_{NO_x}\right)}=0.01} \right)
\tag{5.4}
$$

When integrating over the time the upper interval is assumed to be the constant mean flame front time and not dependent on the equivalence ratio. This is an acceptable approximation for the following reason. The PDF shape is symmetrical around the mean values and a shorter interval time on the lean side is compensated by a longer interval time on the rich side. Even if a theoretical error is added, this effect is not of relevance for practical calculations.

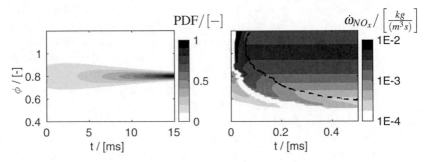

Figure 5.3: Time t and equivalence ratio ϕ dependent data for PDF integration. Left: example of a normalized PDF at $\phi = 0.8$, with an initial unmixedness of 20 %. Right: logarithmically plotted normalized NO_x source term at 8 *bar*, using the same y-axis as left; the dashed line represents the flame front residence time t_{FF}.

The laminar flame front volume (V_{FF}) is calculated by the flame front area multiplied by the thickness. However, the thickness and the area are functions of the equivalence ratio. Therefore, a PDF integration was performed in Eq. 5.5 to account for mixing influence on the thickness. The influence on the flame front area is considered in Eq. (5.6).

$$V_{FF} = A_{FF}\overline{\delta_{l,1D}} = A_{FF} \int_{\phi} \text{PDF}(\phi, t=0)\, \delta_{l,1D}(\phi)\, d\phi \qquad (5.5)$$

For the computation of the flame front NO_x formation rate, the flame front surface was also required. This surface can be computed from the continuity equation by using the flame speed. Hence, the laminar flame speed has to be calculated with the flamelet model in advance. The laminar flame speed dependency on the equivalence ratio is considered in the same manner as the laminar flame thickness for the flame front volume calculation by means of a PDF integration. The described procedure is shown in Eq. (5.6).

$$A_{FF} = \frac{\dot{V}_{FF}\big|_{t=0}}{\overline{S_l}} = \frac{\dot{V}_{FF}\big|_{t=0}}{\int_\phi \mathrm{PDF}\,(\phi, t=0)\, S_l\,(\phi)\,\mathrm{d}\phi} \qquad (5.6)$$

As already discussed, it has been observed that the NO_x formation rate is constant in the post-flame zone. Therefore, the source term was stored just as a function of the equivalence ratio. This is valid if no fuel is induced downstream of the main flames or local quenching takes place. Nevertheless, in the same manner as for the flame front emissions the post-flame NO_x formation rate is obtained by means of a PDF integration multiplied by the post-flame volume in Eq. (5.7). The post-flame volume can be obtained, for example, by means of the mass flow through the combustor divided by the flue gas density multiplied by the post-flame residence time. Another approach could be to estimate the percentage of the post-flame compared to the real combustor volume. This is a reasonable approach for a combustion system with a small flame volume compared to the whole combustion chamber.

$$\begin{aligned}
\dot{m}_{NO_x,PF} &= V_{PF}\,\overline{\dot{\omega}_{NO_x,PF}} \\
&= \frac{V_{PF}}{t_{res} - t_{FF}} \int_{t_{FF}}^{t_{res}} \int_\phi \mathrm{PDF}\,(\phi, t)\, \dot{\omega}_{NO_x,PF}(\phi)\,\mathrm{d}\phi\,\mathrm{d}t
\end{aligned} \qquad (5.7)$$

However, since downstream mixing takes place in practical cases, the change in the PDF over time was considered in a stochastic process. Due to the stochastic process, the PDF multiplied by the source term has to be integrated over the time and the equivalence ratio. Additionally the integral has to be normalized by division through the interval time to get an average formation rate.

The described model takes an equivalence ratio distribution (PDF) as an input to compute a NO_x formation rate, which itself has been translated into a NO_x mass fraction. The optional consideration of a CFD-based mixture distribution (i.e. PDF) as an input is shown in the next subsection.

5.2.3 CFD-based mixture inhomogeneity

The mathematical model from the previous subsection relies on the description of the mixture distribution at the flame front. This distribution can be described based on a PDF, which can be assumed for a hypothetical case or obtained from CFD. In the following, the assumptions for the PDF modeling based on CFD are discussed. This enables the assessment of a combustion system or optimization. All explanations given in this subsection are referring to the application case of Chapter 7.

The source term dependency on the local mixture and therefore on the temperature was modeled by a probability density integration. The PDF is utilized in the previous chapter and can be parameterized by a mean value, the standard deviation and the type of PDF (e.g. Gaussian- or Beta-PDF). However, the location of the sampling surface is also of importance for the PDF to represent the flame front. Within this work, an analytical description has been used to describe the flame front location as shown before [5]. A rotational ellipsoid was used to describe a turbulent Bunsen flame front shape [111].

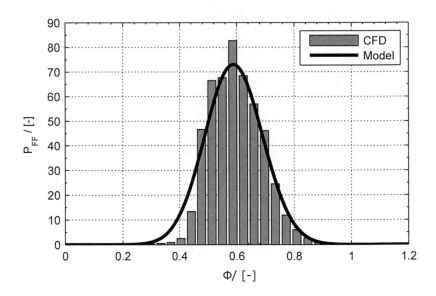

Figure 5.4: CFD-based probability at a Bunsen-type flame front compared to the associated Gaussian PDF

It is assumed that mixture statistics in space and time (spatiotemporal) can be described by a Gaussian type PDF because mixing is a natural process. To check this assumption, sampling was performed from an unsteady non-reacting CFD jet type test case (the same case as in Chapter 7) as sketched in Fig. 5.5. The mixture fraction was sampled on an ellipsoid-shaped flame front for 200 *ms* (after initial 200 *ms*); 5000 time steps and 8448 spatial points from the ellipsoid-shaped sampling plane (as indicated in Fig. 5.5) were considered. The resulting spatiotemporal statistic of more than 42 million data points (5000*x*8448) is shown in the histogram of Fig. 5.4. A similar structure as a Gaussian distributed PDF can been seen. Therefore, the Gaussian PDF obtained by mean value and standard deviation is considered to be a good approximation of the unsteady CFD data.

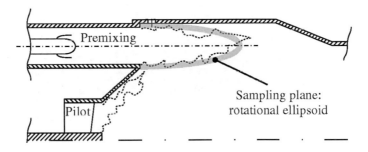

Figure 5.5: Sketch of multi jet burner as sampling environment for the low order approach. Statistics have been obtained by CFD on an artificial rotational ellipsoid.

The mathematical formulation of a Gaussian type PDF is shown in Eq. (5.8) as a function of the equivalence ratio and time. In particular, the time dependency characterizes a stochastic process. Downstream of a flame front further homogenization of the equivalence ratio and therefore of temperature occurs. Therefore, a time-dependent evolution of the PDF was assumed to take mixing into account. The formulation proposed here assumes that the standard deviation changes with time.

$$PDF(\phi,t) = \frac{1}{\sigma(t)\sqrt{2\pi}}e^{-0.5\left(\frac{\phi-\bar{\phi}}{\sigma(t)}\right)^2} \tag{5.8}$$

As described in [81] the evolution of the standard deviation can follow an exponential function. Therefore, Eq. (5.9) has been introduced as a function of a characteristic time. The considered time interval starts at the beginning of the flame front and ends at the outlet of the combustion chamber.

$$\sigma(0 < t < t_{res}) = \sigma|_{t=0}\, exp\left(-\frac{t}{t_{char}}\right) \tag{5.9}$$

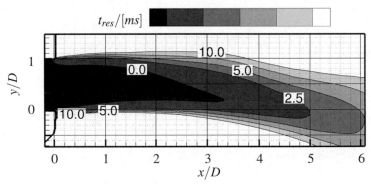

Figure 5.6: Contour plot of mean iso residence time; time starts at a reaction progress of 0.01; the coordinates are normalized with the premixing passage diameter

The model parameter of the previously mentioned stochastic process (i.e. t_{char} of Eq. (5.9)) has to be described. Therefore, the characteristic time (model parameter) was obtained based on a reacting LES of the GT combustion system, described in Chapter 7 and sketched in Fig. 5.5. For this purpose an extra passive time transport equation was added to the LES solver, according to Ghirelli and Leckner [112]. A time transport equation has also been shown by Habisreuther [113]. Eq. (5.10) describes the transport equation of this quantity of residence time, where turbulent diffusion is considered while molecular diffusion is neglected because its amount is significantly smaller for the turbulent flames of interest. The resulting averaged residence time evolution is shown in Fig. 5.6 in the form of a contour plot. Here, the time measurement started at a reaction progress of 1 %. It can be observed that within a period of 5 *ms* of residence time a particle, which travels normally to the flame surface, covers a distance more than ten times farther at the peak of the flame than at the sides. This is due to the high axial jet velocity with small velocity vector components towards the sides of the Bunsen shaped flame.

$$\frac{\partial \overline{\rho}\tilde{\tau}}{\partial t} + \nabla \cdot (\overline{\rho}\tilde{u}_i\tilde{\tau}) = -\nabla \cdot \left(\frac{\overline{\mu}_{sgs}}{Sc_{sgs}} \nabla \tilde{\tau} \right) + \overline{\rho} \tag{5.10}$$

The evolution of the unmixedness over the residence time is shown in Fig. 5.7. The CFD data were obtained at the time-iso surfaces. The model is generated via a method of least squares (regression) and leads to a characteristic time of $t_{char} = 7.1$ *ms*. For a general comparison of the obtained characteristic time, the same curve fit, applied to the published data of Holdemann et al. [75] and Schneiders et al. [76], results in a characteristic time of $t_{char} = 1.75$ *ms* and $t_{char} = 2.53$ *ms*. Those values are of the same order of magnitude, and differences to the present work can be explained by different characteristic diameters or length scales.

For transformation to another system, a coefficient of proportionality k can be calculated (Eq. 5.11) to relate the characteristic time to a characteristic diameter [114]. The amount of the coefficient is about $k \approx 25$ when using the jet diameter, the mean jet velocity and $t_{char} = 7.1$ *ms* as references.

$$t_{char} \approx k\frac{d_{char}}{\overline{u}} \tag{5.11}$$

After describing the PDF shape and the time evolution, the initial standard deviation (i.e. unmixedness) at the flame front has to be derived. The spatial standard deviation is defined in Eq. (5.12). The standard deviation is weighted by the mean mass flow because the formed emissions are transported by mass. The differential quantity dA describes an infinitesimal surface element on the flame front area. Mean values over time are indicated by the $\langle \rangle$ operator; they are obtained by the integration over the time.

$$\sigma_{FF,s} = \sqrt{\frac{\int \langle \rho \rangle \langle u \rangle \left(\langle \phi \rangle - \left\langle \frac{\int \rho u \phi \, dA}{\int \rho u dA} \right\rangle \right)^2 dA}{\int \langle \rho \rangle \langle u \rangle \, dA}} \tag{5.12}$$

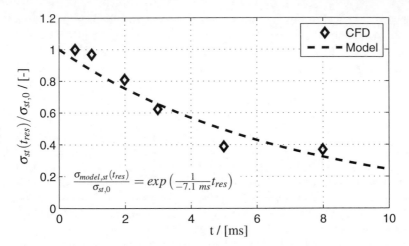

Figure 5.7: Normalized post flame standard deviation vs. residence time; CFD data obtained at instantaneous iso-time surfaces; the model is obtained through regression

To overcome an error due to negligence of temporal effects in the spatial definition of the standard deviation, another variant was introduced. The spatiotemporal standard deviation, Eq. (5.13), accounts for temporal as well as spatial fluctuations in the same manner. This unmixedness can be calculated if the whole data set of the equivalence ratio is available on the flame front in time and space. In comparison to the spatial definition, the spatiotemporal definition is based on precise time-dependent values and integrated over the time.

$$\sigma_{FF,st} = \sqrt{\frac{\iint \rho u \left(\phi - \left\langle \frac{\int \rho u \phi \, dA}{\int \rho u dA} \right\rangle \right)^2 dA dt}{\iint \rho u \, dA dt}} \qquad (5.13)$$

To explore the deviation between both standard deviation definitions, the transient-spatial standard deviation has been defined in Eq. (5.14). The transient-spatial definition in contrast to the spatial definition is based on precise time-dependent data but is not integrated over the time. Consequently, the transient-spatial standard deviation fluctuates with the time.

$$\sigma_{FF,s}(t) = \sqrt{\frac{\int \rho u \left(\phi - \left\langle \frac{\int \rho u \phi \, dA}{\int \rho u dA} \right\rangle \right)^2 dA}{\int \rho u dA}}$$ (5.14)

In combination with the mean equivalence ratio, it is possible to calculate an unmixedness quantity from a flame front standard deviation. The unmixedness is a common quantity to describe the mixing quality [71, 72, 81]. This quantity is a parameter for a description of a Gaussian PDF and on the other hand, it can be used as a target optimization quantity. Depending on the definition of the standard deviation, the spatial, spatiotemporal and transient-spatial definition of an unmixedness quantity can be derived from Eq. (5.15).

$$U = \frac{\sigma_{FF} \int dA}{\int \langle \phi \rangle \, dA} \stackrel{\frown}{=} \sigma_{FF} \frac{1}{\phi_{mean}}$$ (5.15)

An appropriate definition of the standard deviation, i.e. unmixedness, is discussed in more detail in Chapter 7. Also, the influence of non-resolved fluctuations on the standard deviation is discussed.

Chapter 6

Lab-scale validation of CFD & low order approach

6.1 Introduction

As a first validation step for both approaches, lab-scale tests were chosen. Furthermore, a numerical experiment was used for the validation of the CFD-based approach. This synthetic test case was also intended to adjust the model parameter of artificial thickening. The first lab-scale validation was used to determine the model parameter of the turbulence chemistry interaction. Furthermore, a grid study and heat loss investigation were conducted on this case to identify an appropriate choice of grid resolution and setup. The objective of the low order lab-scale validation was to prove the general applicability at perfectly premixed conditions.

6.2 Validation of CFD-based approach

6.2.1 Synthetic laminar test case (1D)

In this section, reference results of 1D premixed prototype flames (using detailed chemistry) are compared to the results obtained by calculating one-dimensional flames with the CFD solver; the flame type was a freely propagating laminar flame (FPLF). The objective of this study is to identify the number of cells needed over the flame front to recalculate the laminar flame speed with a sufficiently small error. This minimum number of cells is an input parameter for the ATF model. For general testing of the tabulation procedure, the laminar flame speed and adiabatic flame temperature are considered over a broad equivalence ratio range.

First, a verification case with an equivalence ratio of $\phi = 0.71$ and a preheat temperature $T_{in} = 573\ K$ was investigated in detail. Fig. 6.1 shows a sketch of the CFD domain. The inlet velocity was selected in accordance with the laminar flame speed of the reference case. The reference simulation was based on the detailed mechanism GRI 3.0. The 1D CFD domain was 20 *mm* long and had a cell size of $0.1\text{x}0.1\text{x}0.1\ mm^3$, while the thermal laminar flame thickness was about 0.4 *mm* according to the reference case.

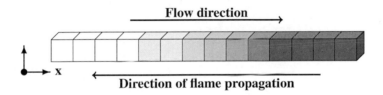

Figure 6.1: Sketch of CFD sub-domain, used to check the reproducibility of the 1D prototype flames by CFD. The color code is an indicator for the temperature. The sub-domain around the flame brush illustrates a flame resolution of approximately six cells.

Table 6.1 shows the percent wise error between the reference solution and the CFD solver as a function of the model parameter n_{Cell}, which describes the resolution for artificial thickening. In the case of an inherent flame thickness above the model parameter, no artificial thickening takes place. After increasing the ATF model parameter above $n_{Cell} = 4$, the thickening factor was proportional to the chosen number of cells. Below and equal to a parameter of four the grid adaptive thickening was inactivated. The calculation of the laminar flame speed for the CFD solver was based on the consumption speed Eq. (6.1). Based on this equation the potential error due to insufficient resolution can be explained: a coarse grid resolution adds a discretization error due to the assumption of $dx \approx \Delta_x$.

$$S_l = \frac{-1}{\rho_{in}Y_{fuel,in}} \int_{-\infty}^{+\infty} \dot{\omega}_{fuel} dx \qquad (6.1)$$

Moreover, table 6.1 shows the maximum value of the thickening factor F_{lam} which is present at 50 % reaction progress according to the model description in the previous chapter. The dynamic thickening model causes a varying thickening factor over the domain with its maximum at 50 % reaction progress. Even for a maximum thickening factor of almost two the laminar flame speed remains unaffected which illustrates a correct implementation of the model.

Table 6.1: Flame resolution, error of the laminar flame speed prediction and the maximum thickening factor as a function of the model parameter. The model parameter describes the minimum number of cells to resolve the flame thickness.

Model Parameter: n_{Cells}	0	2	4	5	6	7	8
Flame resolution	4	4	4	5	6	7	8
ATF parameter: $max(F_{lam})$	1.0	1.0	1.0	1.2	1.4	1.7	1.9
Error: $\Delta S_l / S_l$	8 %	8 %	8 %	4 %	0 %	0 %	0 %

A choice of at least six cells is sufficient to resolve the flame with high accuracy. This observation is in accordance with a previous study [62]. Since the thickening should be as small as possible, $n_{Cell} = 6$ was selected for further work.

The next validation step was to compare flame speed and adiabatic flame temperature between detailed chemistry and the CFD simulations for a wide range of equivalence ratios. The domain was the same as for the investigation of the minimum number of cells, which was coarse enough to cause thickening for all investigated boundary conditions. Fig. 6.2 shows that in the range between $\phi = 0.45 - 1.0$ the deviation from the reference solution is smaller than 5 %. Even for rich conditions with an equivalence ratio from up to $\phi = 2.0$, the temperature was predicted with an accuracy of 5 %. Also the peak location at stoichiometric conditions was reproduced. Hence, the code is valid for lean conditions as well as for partially premixed conditions where locally rich pockets can be predicted. The remaining deviations on the rich side can be explained by the significant contribution of CO and H_2 on this side. As explained for the thermophysical properties, CO and H_2 relied on an analytical approach which slightly overestimated their contribution and resulted in lower temperatures. The velocity deviation may be explained by the resulting error in the transport properties.

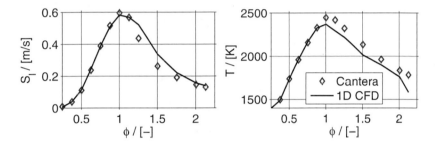

Figure 6.2: Sweep of the equivalence ratio ϕ using ATF ($n_{Cells} \geq 6$); Left: flame speed; Right: adiabatic flame temperature. The diamond symbols indicate the reference solution based on detailed chemistry.

The detailed results for a laminar flat flame are compared in Fig. 6.3 to the prototype flame solution. The diffusion coefficient, in combination with the source term, determines the flame speed. The diagram shows that the mixture average approach works properly for the prediction of molecular diffusion, even if the transport properties for the CFD solver are calculated based on the main species, while the reference solution considers the whole mechanism. This confirms that the diffusion coefficient is determined by the main species of the mixture.

Based on the study presented in this section the thickening parameter was selected to achieve a sufficiently small error regarding flame propagation speed below 1 %. The chosen value of $n_{cells} = 6$ will be used in the following validation and application.

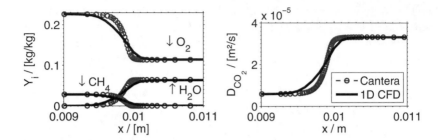

Figure 6.3: Detailed investigation of the $1D$ calculations using ATF from Fig. 6.2 for $\phi = 0.5$ and $p = 10\ bar.$; Left: species distribution; Right: diffusion coefficient.

6.2.2 Laboratory turbulent test case (3D)

For the calibration of the turbulence chemistry interaction model an experiment was selected, which was carried out under gas turbine relevant conditions. The designated experiment used a high pressure (8 bar) triple-jet burner configuration. The perfectly premixed fuel-air mixture was preheated. However, a mixture fraction distribution was present in the combustion chamber due to leakage of air. Non-intrusive laser diagnostic data was available including Raman scattering, $OH*$ chemiluminescence and particle image velocimetry (PIV). The measurements were made at the German aerospace center DLR in Stuttgart [115].

Fig. 6.4 shows the computational domain. The Raman data was sampled on an intersection slice in the x,y-space at $z = -2.4mm$. In comparison to the 20 mm jet diameter, this is a small deviation from the central jet axis. This sampling plane captures the flame and the recirculation zone as well as leakage in air injection. Originally the experiment was not foreseen to consider cold gas injection, but Lammel et al. [115] observed a mixture fraction distribution and they assumed that it was due to air leakage through the optical access seal.

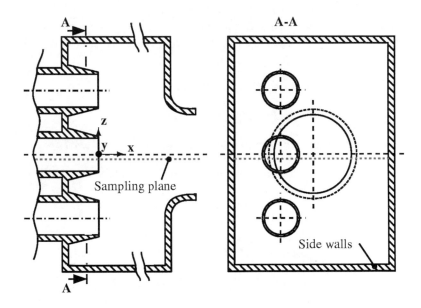

Figure 6.4: Sketch of the triple-jet burner for the computational domain. Left: top view with the three jet openings and coordinate system; Flow goes in the positive x-direction. Right: intersection through inlet jets.

6.2.2.1 Boundary conditions and meshing

In order to perform a validation of practical relevance, the mesh resolution was set to a level which was feasible to achieve for a technical combustion system with moderate computational effort. The computational grid contained 1.4 m. cells and the jets were resolved with 16 cells per diameter (baseline). This choice of refinement was supported by a grid study, which is presented after the results of the baseline setup are discussed. The grid was an isotropic hex dominant mesh with the same refinement level in the flame region as in the jets. In accordance with the 1D study presented in the previous section, thickening was applied to resolve the flame front on a sufficient number of cells (i.e. a model parameter of $n_{Cells} = 6$).

It was assumed that heat loss was relevant because the surrounding windows of the considered case were air-cooled. Proch et al. [101] showed different ways to reproduce this in CFD; he distinguished between the following three cases:

- **No heat loss consideration (adiabatic walls)**

- **Heat loss consideration in energy equation**

- **Heat loss consideration in energy equation and source terms**

The last option considers heat loss already in the prototype flame which is than reflected in the source term of the governing species and controlled by means of an extra controlling variable (β). In the present work, the second option with heat loss modeling on the energy equation and without source term treatment was applied. This choice is supported by a comparative study of the three modeling options at the end of the present section.

Constant wall temperatures combined with boundary layer refinement at the cooled walls were used. Boundary layer refinement is known to be more reliable than using wall functions [116]. The mean non-dimensional wall distance y^+ at the cells next to the chamber wall was about 3. This value is an indicator for the quality of the boundary layer refinement and indicates a resolution of the viscous sub layer by the mesh. The temperature of the side walls was set to 1100 K because the authors of the experimental publication [115] specified the temperature between 920 K and 1200 K. The base plate was set to 600 K because this area was water-cooled. The air leakage temperature was set to 1100 K as the side wall temperature. The leakage mass flow was set to 3 % of the total air mass flow because this is a typical leakage value for this kind of piston ring [1].

The computational domain started five jet diameters upstream of the chamber inlet. In the experiments, there was a bended duct upstream of each inlet jet but to reduce the simulation complexity this was excluded from the simulation domain. In order to mimic the turbulence generated by this upstream geometry accurately, an artificial turbulence generator was used. This turbulence generator worked according to a method proposed by Klein et al. [99]. For the present work an isotropic, turbulent length scale of 1/3 of the jet diameter was selected, in accordance with Klein's results. The selected

---

[1] According to a private communication to the experimentalists from the DLR

turbulent length scale appeared reasonable for a bend. The turbulence intensity was varied until the measured velocity fluctuation at the jet exit could be reproduced. The resulting turbulence intensity at the boundary was at 20 %, which appears to be a high value. However, a strong decay of turbulence in the flow direction has to be considered, especially if the turbulence is artificially created. The domain ended after the exit nozzle, shown in Fig. 6.4.

Table 6.2: Boundary conditions of triple-jet burner for the CFD

fuel	$\dot{m}_{air}$ $[g/s]$	$\dot{m}_{fuel}$ $[g/s]$	T_{in} $[K]$	p_{out} $[bar]$	$m_{leakage}$ $[g/s]$
$CH4$	468.03	19.31	673	8.04	14.04

Table 6.2 gives an overview of the CFD boundary conditions. The look-up tables were generated for pure methane even though natural gas was used in the experiments. The reason was that the detailed composition of the used natural gas was not known. However, methane was judged to be a good surrogate since the combined amount of methane and higher hydrocarbons was specified as $95.3 - 99.5$ % [115].

6.2.2.2 Results

An important step in the present work was to identify the turbulence chemistry interaction model parameter C_Ξ for Eq. (4.21). For this purpose three simulations for different values of the model parameter were performed. It is an established [117] procedure to adjust the turbulence chemistry interaction with an empirical parameter. It was assumed that the model parameter was valid within the validity range of the applied turbulent flame speed correlation, i.e. for GT-relevant conditions.

In Fig. 6.5 the non-dimensional CO_2 concentration is shown for the simulations and the experiment. This is of specific interest since CO_2 was used as the reaction progress variable. The data was sampled on the jet axis of the middle jet and on a transverse line three jet diameters downstream of the jet outlet. The jet axis may be used to assess the flame length, while the transverse line indicates the thickness of the flame brush. The result shows that consideration of SGS wrinkling is important because a parameter of zero

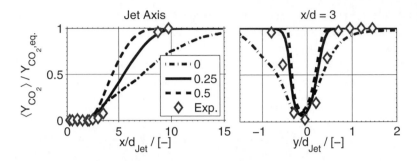

Figure 6.5: Normalized CO_2 mass fraction of the triple-jet burner for the adjustment of the turbulence chemistry interaction. The Legend shows a model parameter C_Ξ (Eq. (4.21)) variation. The results are compared to experiments on the jet axis (left) and transverse to the jet axis (right).

yields a longer flame than observed in the experiment. On the other hand, a model parameter value of 0.5 results in an under prediction of the flame length. It appears that a value of 0.25 for the model parameter gives the best prediction in terms of flame length.

The predicted thickness of the flame brush is slightly smaller than in the experiment, even if the maximum thickening factor is approximately 8 in the flame front. This discrepancy is likely to be due to discrepancies in the velocity fluctuations, caused by a relatively coarse mesh resolution of 16 cells per jet diameter. This assumption was confirmed by a grid study shown in Fig. 6.9.

Fig. 6.6 $c) - f)$ shows a comparison between PIV measurements and simulation results for the model parameter $C_\Xi = 0.25$ with an activated turbulence generator at the inlet. Fig. 6.6 $a), b)$ shows a comparison with deactivated turbulence generator. The modeled SGS fluctuations are indicated by the dashed lines. The velocity profile shows good agreement independently of the turbulence generator. However, the fluctuations were clearly under-predicted when there was no artificial turbulence generation. Using the turbulence generator the fluctuations, i.e. root mean square values (rms) were slightly under-predicted. It is assumed that this deviation was due to a remaining grid discretization error, which added some numerical damping. Nevertheless, the mean and rms profile are a good validation basis for the

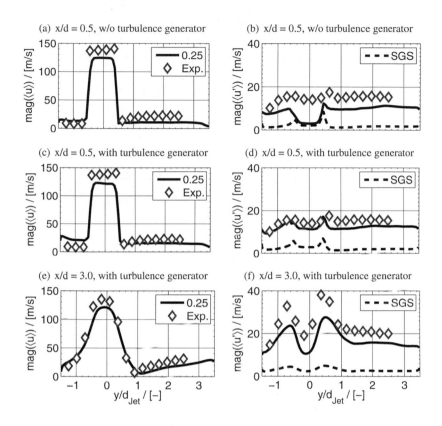

Figure 6.6: Simulated velocity and velocity fluctuation profiles transverse to the jet axis at $x/d = 0.5$ & 3.0 for the model parameter $C_\Xi = 0.25$ (Eq. (4.21)) compared to the experiments (PIV). Left: mean velocity. Right: total velocity fluctuation (solid line) and SGS part of the fluctuation (dashed line). The first row shows results without (w/o) a turbulence generator at the inlet.

LES solver in general.

For the selected model parameter, a detailed comparison of the time averaged CO_2 mol fraction field is shown in Fig. 6.7. The Raman sampling points are indicated by small black dots on the contour plot. The contour itself was generated by triangulation of those points. The overall flame length is comparable to the experiments while the simulated flame brush is a little thinner than in the measurements, as already discussed. When interpreting the results, it should be considered that there were just a few measurement points at the flame tip. Consequently, the experimental field relied on interpolation through a triangulation approach in this region. Therefore, there might be some uncertainty in the flame length prediction. To overcome this problem, the line data from Fig. 6.5 should also be considered.

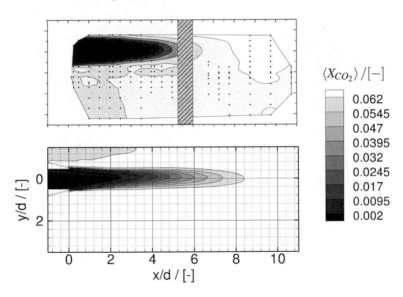

Figure 6.7: Intersection slice in x,y-space with mean CO_2 mol. fraction at $z = -2.4$ mm; top: experiment; bottom: simulation.

A general trend is that in the experiments a non-complete burnout of about 20 % is distributed over the combustion chamber, while the simulation indicates complete burnout at the flame front. This observation must be treated with care since the precision of the CO_2 mol fraction measurement was at about 19 % in the post-flame zone [115]. The experimental uncertainty is of the same magnitude as the difference between the simulation and the experiment. The reduced amount of CO_2 in the top left corner of the simulation is due to the air leakage.

In Fig. 6.8 the standard deviation (rms) of the CO_2 fraction is shown. Comparing the measurements with the simulation results, a good qualitative agreement can be identified. However, the peak values in the flame front were simulated with a broader distribution. As for the previous contour, there was a lack of experimental sampling points in the flame tip area, which may explain the differences in the reaction zone.

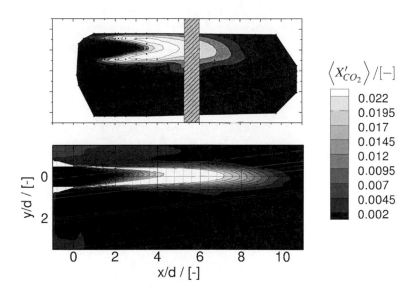

Figure 6.8: Intersection slice in x,y-space with standard deviation of CO_2 mol. fraction at $z = -2.4\ mm$; top: experiment; bottom: simulation.

Based on the presented work, the model parameter for the turbulence chemistry interaction was set to 0.25 because good qualitative and quantitative predictions were achieved. This selection was further tested with the technical application case in Chapter 7. The next two subsections support the choices made for the grid refinement and the heat loss treatment.

6.2.2.3 Grid study

This grid study was performed to justify the chosen refinement level of the lab-scale case but also to provide a strategy for the application cases. Another objective was to confirm a good compromise between accuracy and computational effort. Thus, a significantly coarser grid and a finer grid were compared to the baseline grid. The most important criterion for accuracy from a thermoacoustic point of view is the flame length because this influences the heat release center (i.e. a time lag).

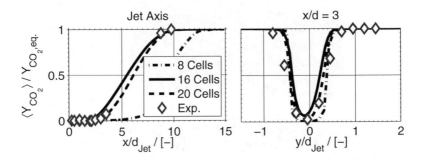

Figure 6.9: CO_2 profiles of the triple-jet burner for different grid resolutions for the model parameter $C_\Xi = 0.25$ (Eq. (4.21)). The Legend shows a variation in Cells per jet diameter. The results are compared to experiments on the jet axis (left) and transverse to the jet axis (right).

Fig. 6.9 shows CO_2 profiles obtained with the different grid refinement level. The choice of 16 cells per diameter as the mesh resolution was justified because the fine grid (20 cells/jet diameter) did not affect the flame length significantly. In contrast, eight cells per diameter would be too coarse and would predict the wrong flame length, which is a crucial parameter for thermoacoustic predictions.

The predicted thickness of the flame brush on the baseline grid is slightly smaller than in the experiment, even if the maximum thickening factor is approximately eight in the flame front. This discrepancy is likely to be due to discrepancies induced by the SGS modeling which is of more importance with coarser meshes. Fig. 6.10 supports this assumption because the total velocity fluctuations in the flame region are comparable between the baseline

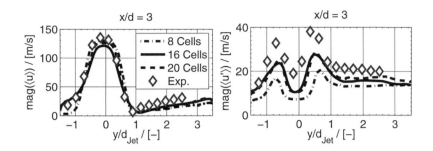

Figure 6.10: Velocity and velocity fluctuation profiles transverse to the jet axis for the model parameter $C_\Xi = 0.25$ (Eq. (4.21)) compared to the experiments (PIV). Left: mean velocity. Right: total velocity fluctuation.

and fine grid configuration. In contrast, the coarse grid results in a different prediction, which assists the conclusion made in the previous paragraph, that eight cells are not appropriate. On the other hand, using a finer grid resolution of more than 20 cells per diameter would not improve the accuracy significantly but would increase the computational costs.

6.2.2.4 Heat loss treatment

In this section the performed heat loss investigation is presented. This investigation shows that the influence on the source term is not relevant for the case considered. Table 6.3 shows the different options of heat loss treatment which have been investigated. The last option, considering the heat loss influence on the prototype flame, consequently affects the source term treatment of the governing species of the CFD.

Fig. 6.11 shows temperature profiles for the different heat loss treatments, while using a fixed model parameter of $C_\Xi = 0.25$. Using adiabatic walls, results in unrealistic high wall temperatures. Therefore, diabatic walls have to be considered as assumed in the boundary description. The influence of the heat loss on the prototype flame resp. source term was not relevant, which is assumed to be due to the high energy density at elevated pressure. For the investigation performed, the heat loss quantity β, defined in Eq. (4.8), was considered in the look-up tables. The heat loss is defined as a normalized

Table 6.3: Haet loss treatment

Legend	Energy eq. in CFD	Heat loss quantity	Prototype flame
adiabatic	Unaffected by walls	$\beta = 1$	Unaffected
diabatic	Affected by wall temp.	$\beta = 1$	Unaffected
diabatic+	Affected by wall temp.	$\beta = $ var.	Affected

defect of total enthalpy. To take the heat loss for the prototype flames into account the enthalpy defect was already considered at the inlet of the prototype flames by reducing the inlet temperature. Therefore, the set of prototype flame calculations was generated for different preheat temperatures, stored in the look-up tables and addressed by the heat loss quantity.

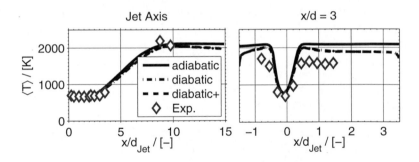

Figure 6.11: Temperature profiles of triple-jet burner for different heat loss considerations, using a model parameter of $C_{\Xi} = 0.25$. Legend:'+' indicates additional heat loss treatment on the source terms. The dashed and the dash-dot lines are overlapping. Results are compared to experiments on the jet axis (left) and transverse to the jet axis (right).

In contrast to the present study, Proch [101] showed an influence on the source term on the flame location, which is assumed to be due to a different pressure level. Proch calculated a jet flame at ambient pressure with a resulting lower energy density, which makes heat loss more relevant in comparison to the overall heat release. Consequently, it is recommended to apply a heat loss model for confined atmospheric test cases.

The remaining difference between measurements and simulated diabatic temperatures may be explained by the assumptions made for the wall temperatures, radiation through the windows and a significantly high measurement uncertainty of 6 % for the temperature profiles [115]. Due to the high uncertainty in temperature, the CO_2 field data used in the previous section is of more relevance. However, this investigation suggests no significant influence of the heat loss on the source term for GT-relevant conditions. Therefore, this has not been considered in the application section.

6.3 Validation of the low order NO_x prediction approach

For validation purposes, a perfectly premixed case from the literature was selected. Biagioli et al. [72, 118] published investigations of a high pressure dump combustor, comparable to the setup sketched in Fig. 6.12. Measured NOx emissions as a function of the adiabatic flame temperature were extracted from the publication. The corresponding boundary conditions are described in Table 6.4. Also a comparison to simulations by Biagiloi et al. [72] is made in this subsection.

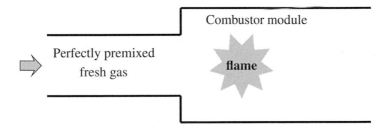

Figure 6.12: Schematic illustration of dump combustor as explained in the literature [118]. The flame is stabilized by the jump in cross-section area and surrounded by adiabatic walls.

Table 6.4: Boundary conditions for the low order validation case based on literature [72, 118]

Boundary condition	Value	Unit
Inflow temperature	723	K
Pressure	20	bar
Air mass flow	0.25	kg/s
Mixture fraction	0.03	–
Volume of flame pipe	896	cm^3
Walls	adiabatic	

The result of the validation study is presented in the left plot of Fig. 6.13. The measured NO_x emissions are up to 50 % below the modeled data but in the range below 1900 K the mismatch is less than +2 $ppmv$. The main mismatch is assumed to be caused by the assumptions, which were made to achieve the required boundary conditions, and by the remaining heat loss. With higher temperatures, the difference between the flame front and total emissions increases due to the higher activation energy needed for the NO_x formation in the post-flame. It is concluded that the model over-predicts NO_x but captures the temperature exponent of the NO_x formation well.

In comparison to the model presented by Biagioli et al. [72], the predicted emissions from the model, introduced in Chapter 5, are higher. The difference between the two models may have two reasons; the uncertainty in the extracted boundary conditions, and differences in the model formulation. However, both these cases neglect the heat loss. Biagioli et al. calculated an integral flame front source term based on a flame front cut off at 99 % conversion of the fuel-air mixture into products. In contrast, the model used in this work accounts for flame front NO_x as long as there is a significant change in the source term. Therefore, the flame front emissions are higher than in the "Biagioli approach" (see Fig. 6.13). Furthermore, the modeling approach of the present work relies on a post-flame source term from freely propagating prototype flame calculations, while the reference approach relies on source terms from plug flow reactor calculations. This explains the remaining difference in the diagram on the right in Fig. 6.13. The underlying work considers the physical effects with more accuracy, and the difference to the experiments is not expected to be due to modeling uncertainties. Rather, there is uncer-

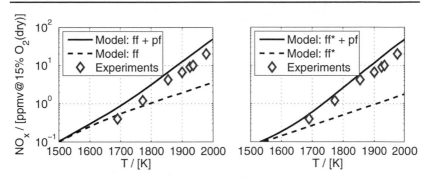

Figure 6.13: Calculated emissions vs. experimental data from Biagioli et
al. [72]. Left: model of the present work; right: mixed model, using a flame
front model of Biagioli et al. [72] and the model of the present work for the
post flame as explained in Table 6.5.

tainty about the experimental measurements, the detailed chemistry reaction
mechanism, and the boundary conditions.

Table 6.5: Legend of Fig. 6.13

Abbreviation	Explanation	Origin
ff	flame front	This work
ff*	flame front	Biagioli et al. [72]
pf	post flame	This work

Chapter 7

Full-scale validation of CFD & low order approach

7.1 Introduction

This chapter is dedicated to the validation of the methods proposed in Chapter 4 and Chapter 5 at full-scale conditions. For this purpose a GT can-annular combustion system was selected. It is a full scale prototype system which was operated at high pressure. For diagnostics it was equipped with emissions and dynamic pressure sensor measurement techniques. Therefore, this system was suitable as a validation case for demonstrating the applicability of the proposed methods and to derive recommendations for the design process.

The most important reason for choosing this test case was to represent real gas turbines. Hence, different physical behaviors were compared between experiments and the simulations at different operational conditions. One objective of this work was to predict emissions. Secondly, the prediction thermoacoustic instabilities were also considered. The location of the flame could be compared to the experiments only indirectly. The previous chapters have already described good accuracy in the flame location prediction. Another interesting aspect of the selected validation case was its composition in different flame types: multi-jet flames on the outer diameter and a central swirl type pilot. The applicability of the proposed methods to the combination of both types of flames could be shown.

This chapter is split into three sections. First, the experimental setup, describing the test rig and the selected test points, is presented. Thereafter, the CFD-based method will be validated in two subsections to predict emissions and acoustics. At the end, a section describes the validation of the proposed low order approach to predict NO_x emissions as a function of the premixing quality. This approach enables to describe the potential of NO_x reduction in a technical combustion system. In contrast, the CFD approach gives more insight at a certain point but is also more expensive. It will be shown that CFD enables multiple physical effects of a single load point to be predicted at the same time, while the strength of the low order approach is to evaluate NO_x emissions over a wide operational range or to perform optimizations.

7.2 Experimental setup of high pressure multi-jet burner

The test facility at the DLR in Cologne provided the infrastructure for full scale testing at high pressure conditions. Fig. 7.1 shows a flow box, which mimics the mid-frame of a heavy duty Siemens GT. The inlet of the flow box consists of a diffuser, which mimics the compressor outlet. Close to the flow box outlet a vane simulation section represents the first turbine vane row. The air flows into a plenum after the inflow through the diffuser. Thereafter, the air makes a turn and flows into the combustion chamber itself which is sketched in a segment of Fig. 7.2.

The burner consists of a central pilot swirl flame surrounded by multiple main jet flames. The jet flames are stabilized by a recirculation zone with a back flow region on the central burner axis. The pilot as well as the main flames are partially premixed. The premixing passages of the main jet stage can be equipped with different mixing devices, which are of special interest for the NO_x prediction. The pilot flame burns on a cooled pilot cone, which has the shape of a diffuser. To allow a tuning margin regarding low load points, the burner has three fuel stages: The pilot stage and two main stages. The main fuel stages alternate in a circumferential direction. The burner concept is explained in more detail in a patent specification [119].

The assessment of self-excited dynamics requires realistic acoustic boundary conditions. In a real GT, the first turbine stage has a certain reflection coefficient from an acoustic point of view. In the described rig, the vane

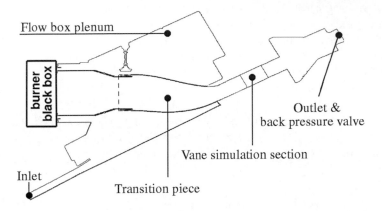

Figure 7.1: Segmental sketch of high pressure test rig with flow box and installed prototype burner.

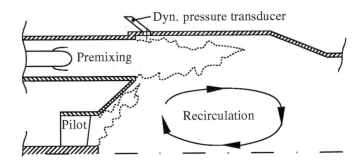

Figure 7.2: Sketch of prototype multi jet burner with outer jet main stage and central swirl pilot.

simulation section is aligned with the axial position of the first turbine row in order to simulate real reflection. However, in contrast to the engine vane section, the simulated vane section does not induce swirl. Downstream of this section a back pressure valve acts like a chocked nozzle. An advantage of the chocked nozzle is that it acts as a non-reflecting boundary condition [120], which simplifies the modeling in the CFD. The burner itself is completely surrounded by the flow box. At the inlet to the burner within the flow box, the cross-section changes abruptly which also provides a well-defined acoustic inlet boundary. The inlet has been approached by means of a reflecting boundary condition. Furthermore, the cross section jump provides a proper turbulent boundary.

For every test point the emission probe signal was averaged over $60\,s$ with a sampling rate of $1\,Hz$. The signal of the dynamic pressure transducer had a frequency resolution of $1.5\,Hz$. As shown in Fig. 7.2, the dynamic pressure transducer was aligned next to the main flames in a branch to consider only static pressure effects close to the heat release center.

For real gas turbine applications the emissions are relevant at the outlet of the whole gas turbine. However, actually the emissions in the first turbine row are already representative for the following reason. The temperature in the first turbine row is reduced due to pressure relaxation and cooling. The reduced temperature makes the NO_x production and CO oxidation so slow that the further change is not relevant anymore. This behavior is caused by the exponential growth of the reaction rate with the temperature according to an Arrhenius formulation. Therefore, the emissions were measured on an intersection slightly between the vane simulation section and the back pressure valve.

An emission measurement area, consisting of 15 geometrically equally distributed flue gas suction probes, was used to sample emissions data. Three water-cooled fingers, each with five suction holes, were aligned on the circumference of this flue gas segment. They were water-cooled to stop further reactions. The openings of the suction holes were in a downstream direction. A gas analyzer form the manufacturer Ecophysics which provided an accuracy of about 5 % [121] was used to analyze the sampled flue gas.

7.3 CFD-based emission and thermoacoustic predictions

The CFD-based approach aims at simultaneous prediction of the key performance parameters of a GT: emissions and thermoacoustics. The method supports the design engineer in a phase where already defined designs have to be assessed. This section describes the full-scale validation of this method in the previously introduced full-scale combustion system. The CFD-based approach is introduced in Chapter 4.

Krediet [42] has already shown the applicability of a comparable approach for self-excited acoustic prediction in a combustion system with only a swirl type flame but without emissions prediction. His approach differs to the present method mainly in the combustion model: he did not use tabulated chemistry and flame thickening. Furthermore, Krediet was restricted to CFL numbers of below one while here even the acoustic CFL number was above one. Goevert [122] applied Krediets approach on the same case as investigated in the present work.

The full-scale validation of the CFD method with the prototype combustion system is shown in this section for a broad range of load conditions. CO emissions are more relevant at part load conditions, while NO_x emissions and thermoacoustics are of interest for all load conditions. However, fuel staging concepts are also of interest for operation of a combustion system in the low load condition. Therefore, all those conditions were reflected by six different operating conditions (see table 7.1), which are appropriate to demonstrate the applicability of the CFD-based method over the whole operational range.

7.3.1 Operating conditions

The selection of the experimental test cases was made to reflect a representative spectrum of typical GT operating conditions. Low load CO increase with two different staging schemes was reflected by four test points. An event of pressure dynamics at base load conditions was shown by an acoustically stable and unstable point – differing just in the air mass flow rate. NO_x emissions were relevant for all test points; table 7.1 gives an overview. The table's equivalence ratio refers to the flow box outlet conditions.

An acoustic instability can be caused by the time lag between heat release and velocity oscillation [43]. Table 7.1 shows that the second base load case was considered as unstable, which can be explained by a lower time lag. The two base load cases differ just in the volumetric flow rate which results in a higher velocity in the premixing passage; the stoichiometric conditions were almost constant. The flow rate modifies the time lag between fuel injection and heat release center, which determines the acoustic stability behavior. In this thesis, the wording 'stable' refers to relative pressure amplitudes below 0.5 %. However, a stability criterion (Eq. 7.1) was also used to identify the stability behavior.

Table 7.1: Boundary conditions and measurement results for multi jet-burner experiments and full-scale CFD validation; the star (*) indicates the multiplication with a constant factor for confidentiality

#	Oper-ation[1]	Fuel staging	FR^2 [−]	Φ^* [−]	\<p\> [bar]	p'/\<p\> [%]	X_{CO}^{*} [3] [ppmv]	$X_{NO_x}^{*}$ [3] [ppmv]
1	BL	no	1.00	0.24	3	0.3	1.0	25
2	BL	no	0.88	0.24	3	3.6	0.8	34
3	PL	no	0.89	0.17	8	0.2	0.9	1.5
4	PL	no	0.89	0.16	8	0.4	10	1.3
5	PL	yes	0.89	0.15	8	0.1	0.1	30
6	PL	yes	0.88	0.13	8	0.1	9.8	8.9

To investigate low load operations, the equivalence ratio was reduced until a certain CO limit was reached. It was assumed prior to the experiments that staged operation would extend the low load limit because the active stages would have provided enough activation energy for full CO burnout. This was confirmed in the experiments, as can be seen in Fig. 7.3. In staged operation, the low load limit could be reduced by more than 10 %. Even if the overall adiabatic flame temperature was lower for staged operations, the flame temperature of the fired main stage was higher than for non-staged operations. This is also the reason for the higher NO_x emissions while staging. To

---

[1]Base load (BL); Part load (PL)

[2]Normalized mass flow rate (FR)

[3]The emissions are corrected to dry and 15 % O_2 conditions. The correction is not further mentioned in the following but the correction was always applied.

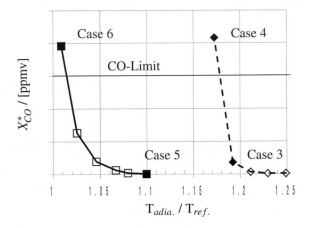

Figure 7.3: CO increase due to low load for different staging schemes; Solid line: Staged main flames; Dashed line: Non-staged

represent the *CO* increase due to temperature reduction, two staged and two non-staged part load test points were selected for the CFD-based validation. Those points were operated at the same pilot equivalence ratio. The only difference was the main stage fuel mass flow.

7.3.2 CFD-setup

In order to predict the acoustic behavior of the system accurately, it must be ensured that the relevant acoustic eigenmodes are located completely inside the computational domain. Therefore, the flow box including the whole burner was selected as the computational domain based on information obtained from previous 1D acoustic simulations [42]. These simulations suggest that the acoustic eigenmode extends from the transition between the flow box plenum and the premixing passages to the pressure control valve at the outlet, which was also later confirmed by the simulation results shown in Fig. 7.6. In this figure the time series of the pressure field was used to create a discrete Fourier transformed (DFT) 3D field of the 250 Hz Mode, which showed the eigenmode. The Fourier transformation transforms the pressure signal from the time domain into the frequency domain. After the transformation, the pressure amplitude as well as its phase can be illustrated for certain frequencies.

The CFD setup was created taking into consideration the objectives of flame, thermoacoustics and emissions modeling. Therefore, most of the geometry features were fully resolved by the grid. There were just two regions which were modeled by using separate inlet boundary conditions: purge air at the burner head end plate and cooling at the transition piece between the burner and the vane simulation section. The amount of the non-resolved purge air was at about 1 % of the inlet mass flow. The cooling mass flow of the transition piece of about 10 % is significant but downstream of the combustion chamber itself. Therefore, both artificially modeled inlet boundary conditions were considered as not relevant for the combustion itself. The separately modeled inlet mass flows are subtracted from the flow box inlet mass flow to fulfill the overall mass balance. The location of the cooling injection into the transition piece is modeled by two slots, while in reality the cooling was more distributed over multiple holes.

The different $19 - 24\ m$. cell grids are consisting – by more than 80 % – of hexahedral cells, while the rest were polyhedral cells. The numerical settings were the same. Even though the focus of the part load investigation was not on acoustics, the domain contained the same details as for the base load investigation. The experimental hardware varied in the different cross-section areas of the exit nozzle (i.e. pressure levels). The cell size in the flame region (pilot + main flames) was slightly changed between the base and part load cases from $2\ mm$ to $2.5\ mm$ in the flame region. The other regions changed accordingly. The main reason for the small change in the grid resolution was to use the baseline grid strategy (about 16 cells per jet diameter) for the part load case as proposed for the triple jet burner validation case in Chapter 6. In contrast, the base load grid was a little finer (about 20 cells per jet diameter) in order to be comparable to previous calculations [122], which are not a concern of this work. Most of the regions were resolved according to a refinement of at least eight cells per characteristic length scale. The exceptions were the pilot cone cooling channel and the high frequency resonator purge air holes with at least four cells per characteristic length.

Boundary layers were modeled by using so-called wall models. De Villiers [116] describes the usage of wall models in the OpenFOAM environment. His recommendation is to use a normalized wall distance of at least

$y^+ = 40$. Without using wall models, y^+ should be about one. This would have increased the number of cells drastically. Therefore, wall models were used in this work and the advice of de Villiers was considered in most of the regions.

The outlet boundary condition was set to the high Mach side of the back pressure valve as shown in Fig. 7.1. From this location no acoustic wave could travel upstream. Thus, the acoustic properties of the specified outlet boundary had no impact on the simulation results and a non-reflecting boundary condition could be used. On the upstream side of the combustion chamber, the inlet boundary condition was set upstream to the inlet of the flow box plenum. Since the open area of the inlet was small compared to the plenum wall surface, using a reflecting boundary condition is assumed to have introduced little error. This reflecting type of boundary was used for reasons of simulation robustness.

For the selection of the time step, the acoustic CFL number, which was defined in Eq. (4.33) in Chapter 4, has to be considered. The acoustic CFL number includes the convective velocity and the speed of sound to consider for acoustic waves propagating through the domain. A maximum acoustic CFL number of 2 was chosen in order to exclude numerical dispersion [42]. Assuming a combined speed of 1000 m/s, a cell width of 2.0 mm and a maximum acoustic CFL number of 2.0, results in a time step of $4e-6$ s for the designated full-scale validation cases.

In Chapter 4 a separate study for the assessment of the numerical error was performed. Assuming a longitudinal 250 Hz wave in the test rig, a resolution of about 1000 cells per wave length can be assumed based on the cell size. For the estimation of the cell size, the coarse cells downstream of the flame were considered as representative of the whole domain. Fig. 4.4 shows that the numerical error (Eq. (4.34)) was at about 0.01 for this resolution and an acoustic CFL number of 2.0.

The numerical error of about 0.01 was smaller than the growth rate observed in the experiments (about 5 % amplitude change per wave length). After reaching a limit cycle, the amplification was in equilibrium with the physical damping [42]. Thus, also the physical damping – of at least 0.05 – was sufficiently larger than the numerical damping. The whole domain contained about one to two longitudinal wavelengths, so that the effect of the numerical error would be minor anyway. This can be stated because the influence of the numerical error is considered as a relative error per wavelength.

The numerical settings changed within the first 10 *ms* from a first order time discretization to the blend between the first and second order approach which was described in Chapter 4. It is assumed that changing the numerical schemes provided enough initial disturbances to trigger self-excited dynamics.

7.3.3 Thermoacoustic prediction

Previous studies were focused on either emissions or thermoacoustic stability predictions, without connecting them. In contrast, the method proposed in Chapter 4 is used in this study to predict both at the same time. This is possible by using tabulated chemistry for different zones and a fully compressible CFD approach. As a result, one can see a NO_x emission increase with high thermoacoustic instabilities later in Fig. 7.14. However, this subsection is mainly focused on the crossover from stable to unstable acoustic conditions at base load.

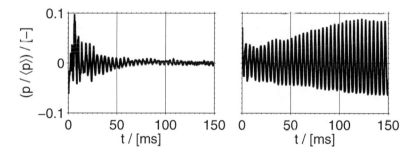

Figure 7.4: Pressure signal at the position of the dynamic pressure transducer for base load conditions for thermoacoustically stable (left) and unstable (right) conditions

First, the pressure signal of the unstable base load case was investigated to work out when the limit cycle was reached. Fig. 7.4 shows an increase of the pressure amplitude over the first 100 *ms*. Thereafter, the limit cycle is reached, indicated by constant pressure amplitudes. Based on this observations, averaging was performed after 100 *ms* for both base load cases. For comparison, the mean residence time in the combustion chamber itself was at about 30 *ms*. The averaging time was 15 *ms*, which included three periods of a 250 *Hz* wave. The initial perturbation was due to the starting solution and the abrupt change to the second order scheme which acted as a Dirac excitation.

The characteristic frequency of 250 *Hz* at unstable conditions was determined by a discrete Fourier transformation (DFT) of the pressure in Fig. 7.5. The evaluation in the frequency space illustrates further characteristic frequencies; they are multiples of the 250 *Hz* mode. Comparing the CFD-based signal with the experiments, confirms the reliability of the frequency predictions. The simulated frequency spectra tend to have slightly higher amplitudes but the characteristic frequency is predicted well. The over prediction of amplitudes may be explained by an over- prediction of the excitation or under-prediction of the physical damping mechanisms. Errors due to the boundary condition formulation can be excluded since the mode is fully contained inside the computational domain (see Fig. 7.6). A potential damping mechanism, which was not included in the simulation, is the transition cooling system. However, the relevance of this mechanism has not been quantified yet. On the excitation side, the combustion model may inaccurately predict the dynamic flame behavior, but this cause is less likely, since the same over-prediction of pressure amplitude for the same case was observed by using a turbulent burning velocity type combustion model [122].

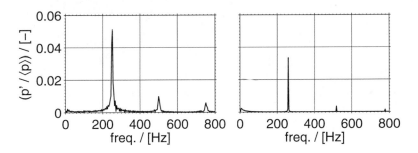

Figure 7.5: DFT of pressure signal at the position of the dynamic pressure sensor for thermoacoustically unstable conditions; Left: simulation, Right: experiment.

Fig. 7.6 confirms the proper choice of the boundary conditions; the fundamental acoustic mode is fully contained inside the combustion chamber. One wave length is contained in the area between the premixing passages and the back pressure valve. The phase plot (Fig. 7.7) illustrates the standing wave character of the mode. The phase is constant over a wide range and changes rapidly from positive to negative values at the position where a node of the acoustic 250 Hz mode is located.

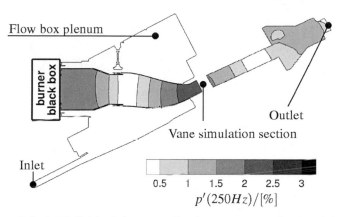

Figure 7.6: DFT field of the normalized pressure amplitude of the 250 Hz mode at unstable acoustic conditions (case 2).

In the acoustically stable base load case the pressure amplitude decreases from a maximum of about 10 % to below 1 %, as can be seen in Fig. 7.4. The limit cycle of the unstable case is reached at about 6 − 8 % pressure amplitude. It seems that there is a small drift of the mean pressure towards higher pressure values because the positive amplitude is higher than the negative. This might be due to an increasing pressure loss due to acoustic induced turbulence. Due to the high computational costs, this effect has not been investigated further with a longer run time, but it is assumed that the results would not be significantly influenced since the pressure drift is small. The effect of a pressure drift is discussed in more detail in the emissions prediction section because it may influence the mixture fraction.

The development of the limit cycle amplitude can also be observed in the heat release time series in Fig. 7.8. The heat release is obtained by a volume

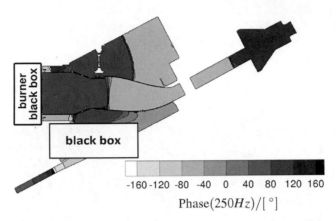

$$\text{Phase}(250Hz)/[\,^{\circ}]$$

Figure 7.7: Phase of the 250 Hz mode at unstable acoustic conditions (case 2). The detailed description of the black box is excluded for confidentiality reasons.

integral of the normalized CO_2 source term over the whole combustion chamber. Assuming a one-step reaction, the CO_2 source term is proportional to the fuel consumption and in consequence to the heat release rate.

For a general stability analysis a Rayleigh criterion [39] can be evaluated. The original formulation of the Rayleigh criterion is presented in Chapter 3. In Eq. (7.1) the criterion is defined as the integral of the heat release fluctuation multiplied by the pressure fluctuation at the dynamic pressure transducer. If these fluctuations are in phase, the criterion is positive. If the system gets damped, the value is negative or tends to zero because the fluctuations vanish in stable conditions. Fig. 7.9 shows the Rayleigh criterion as a function of the time. The criterion was based on an integration length of one period τ of a 250 Hz mode; the pressure signal of Fig. 7.4 and the heat release of Fig. 7.8 were used as inputs for this evaluation. In summary, it can be stated that the Rayleigh criterion is positive for the unstable case as expected for unstable conditions. In contrast, the stable case shows a Rayleigh criterion of zero which can be expected because the pressure amplitudes are in the range of noise.

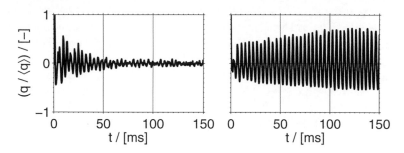

Figure 7.8: Predicted time series of heat release for thermoacoustically stable (left) and unstable (right) base load operating conditions. The heat release signal was obtained by transient integration of the heat release over the whole domain volume.

$$RC = \frac{1}{\tau} \sum_{t-\tau}^{t} q' p' \Delta t \qquad (7.1)$$

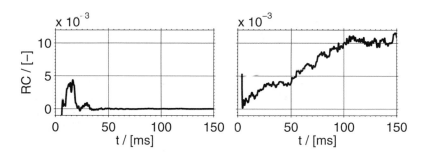

Figure 7.9: Rayleigh criterion for thermoacoustically stable (left) and unstable (right) conditions.

The experimental and simulated pressure spectra at acoustically stable base load conditions are shown in Fig. 7.10. As for the unstable case, the frequency of the fundamental eigenmode was predicted very well and the acoustic pressure amplitude was slightly over-predicted. Moreover, the simulation shows multiples of the characteristic frequency which were not present in the experiments. The reason is assumed to be the same as for the acoustically unstable conditions; probably the damping mechanism of the simulation was slightly too small. Nevertheless, the amplitudes are very small in comparison to Fig. 7.5. The results are very promising because the small amplitudes suggesting a acoustically stable behavior as already explained by means of the Rayleigh criterion. The peak at frequencies close to 0 Hz is caused by white noise, which was created by a discrete length and resolution of the time signal.

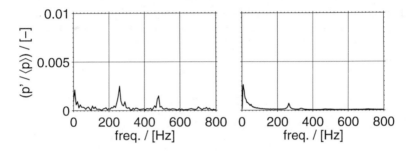

Figure 7.10: DFT of pressure signal at the position of the dynamic pressure transducer for thermoacoustically stable conditions; Left: simulation, Right: experiment.

At part load conditions no significant thermoacoustic instabilities were observed in the experiments. However, the modeled pressure spectra are shown in a zoomed manner in Fig. 7.11 to investigate the expressiveness of the model. Firstly, no high amplitudes were observed in the modeled spectra in comparison to case 2. However, secondly, the characteristic frequency of a 220 Hz mode could be observed in non-staged operations. This frequency was slightly lower than at base load because the mean temperature in the combustion chamber was about 25 % lower than in base load conditions; this resulted in a lower speed of sound, which is proportional to the frequency. In summary, even acoustically stable part load calculations are appropriate to get an idea of the characteristic frequencies of a combustion chamber.

To deduce the frequency at base load the part load frequency should be

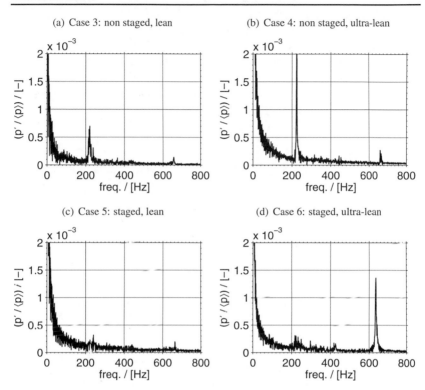

Figure 7.11: Simulated DFT of pressure signal at the position of the dynamic pressure transducer at part load conditions. The experimental pressure amplitudes were below 0.5%

multiplied by the square root[1] of the temperature ratio. At staged conditions, the amplitudes of 220 Hz almost vanished, which was probably due to damping by the non-fired stage. At ultra lean staged conditions (case 6) higher frequency dynamics occur; there might be a connection to lean blow off because the operational mode was close to lean blow off. Anyway, the amplitude was still far below the amplitude of case 2 and can be considered as stable.

---

[1]Because the speed of sound is calculated as follows: $c = \sqrt{\kappa R T}$

7.3.4 Emission prediction

An objective of this work was to predict the CO and NO_x emissions behavior at different load conditions. However, the influence of different staging schemes was also investigated. As shown in Fig. 7.3 a CO increase with leaner conditions could be observed in the experiments, while staging extended this range. However, due to hotter peak temperatures at staged conditions, a NO_x increase was observed as well.

Table 7.2: Experimental CO emissions compared to simulation results.

#	Fuel staging	Exp. X_{CO}^* $[ppmv]$	Sim. X_{CO}^* $[ppmv]$	ΔZ_{error} $[\%]$	ΔT_{error} $[K]$
1	no	1.0	3.0	+0.1	+1
2	no	0.8	3.2	+0.4	+3
3	no	0.9	0.4	−0.5	−4
4	no	10	0.3	+2.2	+14
5	yes	0.1	0.6	+4.2	+26
6	yes	9.8	5.2	+2.8	+16

Table 7.2 shows the CO simulation results in comparison to the experiments as well as the discrepancies in the prediction. At base load conditions the results are reasonable. The emissions are considerably below the CO limit and the CO limit may determine the operational window of the GT. The base load emissions are at equilibrium and the remaining error may be explained by measurement uncertainties at those small concentrations.

At staged operations a significant CO increase due to lean operation could be simulated even if the absolute value at case 6 was off. The error in the prediction was probably due to a deviation in the exit temperature, as also shown in the table. The temperature deviation was calculated based on the mixture fraction difference. The root cause of this deviation is explained in the next paragraph. An estimation of the error influence can be made based on the left CO curve of Fig. 2.3, assuming that the CO increase at part load is kinetically driven. In this figure, the amount of CO is approximately doubled with a decrease of $10\ K$. Considering this, the CO increase would be even more significant. Nonetheless, the CO increase at non-staged part load operation conditions could not be simulated. Considering the temperature offset, the prediction would be more reasonable. However, due to the steepness of the kinetically driven CO increase, small uncertainties in the local mixture fraction or the temperature result in large uncertainties in the CO prediction. Therefore, the mixture fraction at the point of interest has to be monitored with care.

$$\frac{\mathrm{d}m}{\mathrm{d}t} = \frac{\mathrm{d}\rho}{\mathrm{d}t}V = \frac{\mathrm{d}p}{\mathrm{d}t}\frac{R}{M}TV \qquad (7.2)$$

As mentioned in the previous paragraph and shown in table 7.2, there was a discrepancy in the temperature prediction. As also mentioned the temperature differences were calculated based on the mixture fractions. The temperatures were calculated by using the GRI 3.0 mechanism to obtain equilibrium temperatures. The mixture fraction deviation can be explained by compressible flow effects. The initial pressure and in consequence the initial density were off in the part load simulations. Eq. 7.2 explains a change of mass in a volume due to a density change. Assuming ideal gas, the density gradient is proportional to a pressure rate. Due to the large flow box volume of approximately 1 m^3, a pressure gradient can indicate an accumulation or a reduction of mass as a slow transient process. In contrast to the air, the fuel is not contained in a huge volume before it enters the premixing passages. Hence, a slow pressure rate affects the air mass flow, while the fuel mass flow can be assumed to be unaffected. Consequently, the proportions (i.e. the mixture fraction) are affected.

Fig. 7.12 shows the change of pressure with time. A positive rate results in an air mass accumulation and consequently in a mixture fraction increase; for a negative gradient and vice versa. Comparing table 7.2 with the Fig. 7.12 confirms this statement. Unfortunately, very long run times would be necessary to make the pressure gradient negligible. Even after 600 ms there was a remaining gradient but a longer run time would have exceeded the budget for the calculations. Those issues may be mitigated by choosing better initial pressure values or by mapping from a solution, obtained on a coarse grid. The short lag of data at about 320 ms in Fig. 7.12 d) is due to a manual mistake where sampling data was overwritten, but due to the short period no significant influence on the DFT spectra is expected.

To gain a detailed insight into the successfully modeled CO increase at staged conditions, post-flame progress contour plots of case 5 and 6 are shown in Fig. 7.13. The figure shows that the last percentage of progress determines

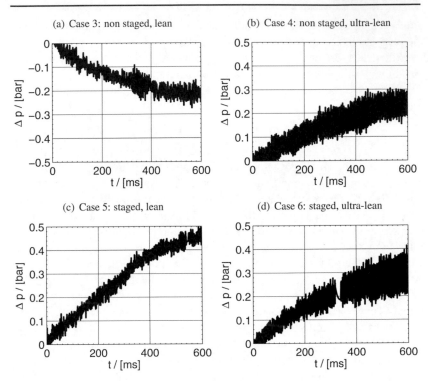

Figure 7.12: Simulated pressure signal drift of the part load cases at the location of the dynamic pressure transducer over the last 600 *ms*.

the amount of CO. For case 5, equilibrium is reached in the combustion chamber itself, while the process is significantly slower for the ultra-lean conditions. This figure also suggests that even a small equivalence ratio or temperature changes can cause high emissions. Indeed, the part load CO emissions are determined by the exponentially decaying last percentage of the CO oxidation. This illustrates the necessity of having a separate post-flame progress variable.

As for the CO emissions, a comparison between experimental and simulated NO_x emissions is shown in table 7.3. The following general trends were predicted very well: a NO_x increase towards base load, an increase as

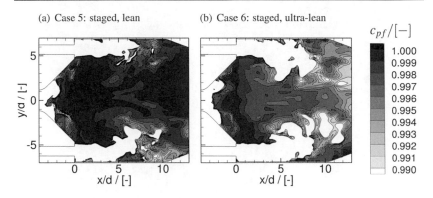

Figure 7.13: Reaction progress of post-flame c_{pf} for lean and ultra-lean conditions.

a result of staging, a decrease with leaner conditions for staged operations and an increase of NO_x emissions at base load and in unstable acoustic conditions. In most cases the absolute values were under-predicted. This was not expected based on the over-prediction of the outlet temperature but may have been caused by artificially faster fuel air mixing than in reality. A measurement uncertainty of this order is rather unlikely. Artificial mixing can be caused by coarse grids and transitions between grid sizes because of numerical errors. In particular, the rather fast decay of the NO_x at the beginning of the post-flame might not be resolved properly.

For a detailed insight into the NO_x production, the increase at base load is investigated in more detail. Two mechanisms caused the NO_x increase at the unstable conditions. Firstly, the thermoacoustic instability was triggered by a reduced volume flow; consequently, the residence time was increased, which increased the post-flame emissions. However, secondly also mixture fraction fluctuations, induced by the acoustic instabilities, caused an increase, as can be seen in Fig. 7.14 and Fig. 7.15. At the position of high fluctuations, the highest emissions were present.

The fluctuation, i.e. standard deviation of the mixture fraction[2], was

---

[2]$Z_{rms} = \sqrt{\frac{1}{t}\sum_{t=0}^{t}(Z(t) - \langle Z \rangle)^2 \Delta t}$

Table 7.3: Experimental NO_x emissions compared to simulation results.

#	Fuel staging	Exp. X_{NO}^* [$ppmv$]	Sim. X_{NO}^* [$ppmv$]	ΔZ_{error} [%]	ΔT_{error} [K]
1	no	25	14.9	+0.1	+1
2	no	34	22.6	+0.4	+3
3	no	1.5	1.4	−0.5	−4
4	no	1.3	1.8	+2.2	+14
5	yes	30	10.8	+4.2	+26
6	yes	8.9	4.7	+2.8	+16

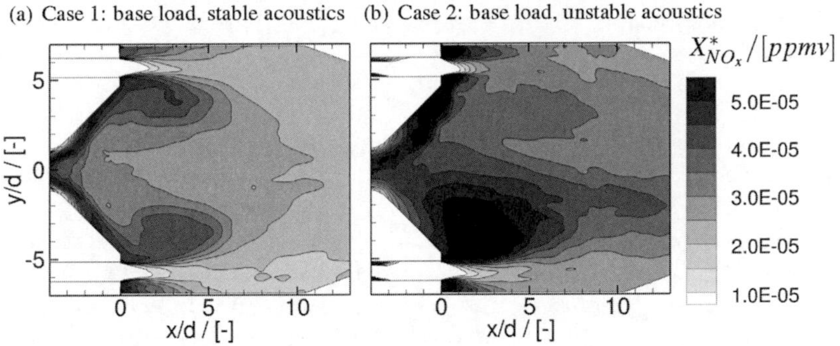

(a) Case 1: base load, stable acoustics   (b) Case 2: base load, unstable acoustics

Figure 7.14: Mean NO_x Emissions for base load conditions for thermoacoustically stable (left) and unstable (right) conditions

caused by the acoustic instability. This observation is justified because only the flow rate was changed between both cases, but the mean mixture fraction was kept constant. Actually, a reduced flow rate is assumed to lower temporal fluctuations rather than increase them because turbulence is reduced. Fluctuations at unstable conditions may also be explained by a thermoacoustic feedback loop. The feedback loop is caused by a density change in a flame due to a varying mixture fraction. This density change provokes a pressure wave which modifies the fuel injection. The variation in fuel injection is the reason for the density change in the flame and closes the feedback loop. Based on the comparison between Fig. 7.14 and Fig. 7.15 it can be stated that modeling of the interaction between thermoacoustic effects and emissions was demonstrated.

(a) Case 1: base load, stable acoustics   (b) Case 2: base load, unstable acoustics

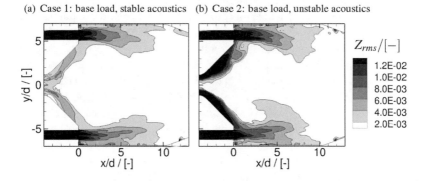

Figure 7.15: Standard deviation of mixture fraction Z_{rms} for base load conditions for thermoacoustically stable (left) and unstable (right) conditions

7.4 Low order prediction of mixing influence on NO_x emissions

The low order approach aims at parametric studies of the operational conditions and of the geometry in order to assess the influence of these parameters on the NO_x emissions. In comparison, the CFD-based approach leads to a comprehensive understanding of a certain configuration at certain boundary conditions. The low order method supports the design engineer in an early design phase to determine the fundamental parameters of a combustion system, e.g. the residence time and the required mixing quality. This method is used in this section to assess the NO_x potential of the previously introduced experimental setup of the prototype combustion system in terms of mixing and residence time at base load conditions. Unfortunately the cases from the previous subsection of the CFD-based assessment could not be used for the validation of the low order approach because for those cases the pilot flame determined the NO_x emissions. In contrast, the low order approach relies on the prediction of the main flames. Therefore, other test cases without major pilot flame contribution were selected.

The fundamentals of this approach are described in Chapter 5. They are based on a former publication [5] and a comparable concept, presented before [72]. For the assessment of the mixing quality of different configurations an unsteady RANS approach was used according to the description in the former publication [5]. This approach differs from the previously applied LES approach mainly by the turbulence modeling: the unsteady RANS approach relied on a $k - \omega$ two equation turbulence model. Even if the unsteady RANS approach is less accurate, large scale turbulent effects were resolved [87]. The main reason for using the less accurate approach for a mixing assessment was that it is a method which is cheap enough to perform optimization runs.

For the purpose of validation, base load emission measurements with different premixing configurations were carried out, after the configurations were investigated regarding their mixing quality. This section describes the evaluation of the mixture quality based on CFD and their NO_x assessment via the low order approach. Afterwards, the resulting emissions were compared to the measurements and a fundamental investigation of the NO_x emissions was performed as a function of the mixing quality and the residence time.

7.4.1 Boundary conditions

As mentioned before, the low order method was applied to assess the NO_x potential of the presented prototype combustion system at base load conditions. Therefore, certain premixing systems were evaluated by transient CFD calculations in advance. Two fundamentally different fuel injection concepts were compared via CFD in different variants by their mixing quality.

Table 7.4: Boundary conditions and experimental results for multi jet-burner (see Fig. 7.2) premixing evaluation

#	Operation	Pressure	Dynamics	Mixing[1]	Exp. $X^*_{NO_x}$
7	base load	8 bar	stable	JIC	23 ppmv
8	base load	8 bar	stable	JIC+	23 ppmv
9	base load	8 bar	stable	VG	15 ppmv
10	base load	8 bar	stable	VG+	14 ppmv

Table 7.4 shows the differences between the test points which were first evaluated regarding their mixing quality by CFD and then investigated regarding their emissions in the experimental setup. The jet in cross-flow (JIC) concept consists of one or more fuel gas jets, which were injected perpendicular to the air flow. Further details are described in [77]. The vortex generators (VG) were applied to add artificial turbulence, which enhances the mixing quality. The '+' sign in the Mixing column of the table indicates an enhanced variant of the mixing concept.

The computational domain was significantly reduced and primarily included the premixing system. The mesh was an unstructured hex dominated mesh as used before. In contrast to the LES approach, only a segment of the burner was meshed to keep the computational costs as low as possible. The refinement of the grid was at about 30 cells per jet diameter at the exit of the premixing passage. This was a finer resolution than for the LES. The fine grid was used to partially overcompensate for the lower accuracy of the unsteady RANS simulations. The CFD setup as well as the grid strategy were validated on a cold flow jet in cross-flow measurements in advance.

[1] Jet in crossflow (JIC); Vortex generators (VG)

7.4.2 Mixing evaluation

For the assessment of the different premixing systems, the mixing quality (i.e. unmixedness) was predicted by non-reacting, unsteady RANS simulations. A possible future application of the mixing assessment could be an optimization run; a CFD-based target optimization quantity could be used. As a possible target optimization quantity different definitions of unmixedness parameters were proposed in the method description of Chapter 5. Steady simulations can provide spatial unmixedness quantities and a spatiotemporal unmixedness can be obtained based on unsteady simulations. Both quantities are investigated in this subsection by means of the different premixing variants.

To account for unsteady effects while reaching computational efficiency, non-reacting unsteady RANS simulations were used. In this context an implicit discretization method allowed for CFL numbers above one. The domain's maximum CFL number of about 80 was observed in the fuel nozzles of the premixing passages. This CFL number is about twenty times higher than observed for the LES approach and explains the gain in computational efficiency. Nevertheless, the CFL number is about three in the region where a flame would be expected. This would be still too high to achieve reacting LES calculations, because the species equations were solved explicitly in the LES. The $k\omega$-SST model was chosen for turbulence modeling and next to the wall, a wall model was used.

Table 7.5: Boundaries and calculated unmixedness values of for the full-scale validation (see Fig. 7.2) of the low order approach.

#	Operation	Pressure	Dynamics	Mixing	U_{st}	U_s
7	base load	8 *bar*	stable	JIC	28 %	11 %
8	base load	8 *bar*	stable	JIC+	29 %	10 %
9	base load	8 *bar*	stable	VG	22 %	7 %
10	base load	8 *bar*	stable	VG+	24 %	13 %

Table 7.5 shows the resulting unmixedness values based on the different formulations. Generally, the spatial unmixedness is smaller than the spatiotemporal unmixedness, which may be explained by the exclusion of temporal effects. Assuming statistically independent spatial and temporal

distributions, a resulting corrected distribution would also be normally distributed [123]. Furthermore, the square of the corrected standard deviation Eq. (7.3) would be the sum of the squares of the spatial and temporal standard deviation. This explains the difference between the spatial and spatiotemporal unmixedness. This explanation can be visualized by plotting the transient spatial unmixedness over time.

$$\sigma_{FF,st}^2 \approx \sigma_{FF,s}^2 + \sigma_{FF,t}^2 \qquad (7.3)$$

In Fig. 7.16, the spatial, spatiotemporal and transient spatial unmixedness, according to their definition in Chapter 5, are shown for the simulation of the VG. It can be observed that the spatial and the transient spatial unmixedness are of the same order of magnitude at the beginning of the sampling at 0.2 *s*. The different unmixedness quantities are diverging with averaging time. The reason is that a time-averaged quantity is unequal to a quantity based on mean values. Indeed, the diagram points out the difference between RANS calculations and transient CFD calculations (e.g. based on LES). The spatial unmixedness was calculated based on the averaged values, using 0.2 *s* as a start time. Therefore, the spatial unmixedness is comparable to an unmixedness obtained by RANS calculations. The time-averaged mean value of the transient spatial unmixedness of 0.2 is of the same order of magnitude as the spatiotemporal unmixedness of 0.22.

Assuming normally distributed fluctuations in time, the spatial standard deviation could be transformed into a spatiotemporal standard deviation by means of Eq. (7.3) if the temporal standard deviation would be known. This could be a method to correct RANS based results.

7.4.3 Results

The CFD-based mixing assessment of the previous subsection is used within this section for further validation purposes. First a general overview of the emissions as a function of two key design parameters – residence time and mixture quality – is shown in a contour plot in Fig. 7.17. Thereafter, the discrete volume resp. the residence time of the already described multi jet burner (see Fig. 7.2 as reference) was considered for the low order calculation of the emissions. The residence time and the thermodynamic boundary conditions as input parameters lead to a NO_x mixing dependency. This low order NO_x model allows for a comparison with the CFD based mixing assessment. The

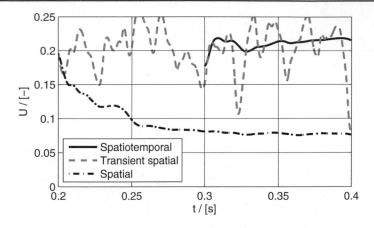

Figure 7.16: Comparison of spatial, spatiotemporal and transient spatial un-mixedness over time for case 9

comparison can be used to find a proper definition of the unmixedness. The following two types of unmixedness quantities are discussed:

- **Spatial unmixedness**

- **Spatiotemporal unmixedness**

Fig. 7.17 explains that reducing the residence time and improving the mixing quality results in a NO_x emissions reduction. However, when reducing the residence time, the negative impact on the CO turn-down behavior has to be considered. Above 10 ms residence time and 10 % unmixedness, the emissions are increasing rapidly with the unmixedness; but with a better mixing quality the influence of the unmixedness decays, while the influence of residence time appears linear. Below an unmixedness of about 10 % the effect on residence time reduction seems to be more reasonable from a NO_x emissions reduction perspective. Nevertheless, for a quantitative assessment, the type of unmixedness needs to be defined.

In Fig. 7.18, the experimentally measured NO_x emissions are compared to two different unmixedness definitions. The unmixedness was obtained by the CFD-based mixing evaluation for the corresponding cases. Moreover, the modeled NO_x emissions, using the low order model, are shown in the plot as

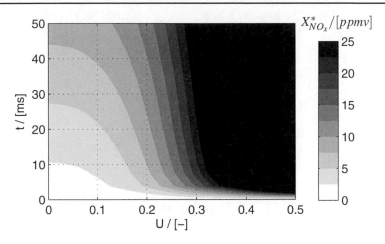

Figure 7.17: Predicted NO_x emissions as a function of unmixedness and residence time

lines. These modeled emissions are plotted over the flame front unmixedness to assess the different definitions of unmixedness.

It can be seen that there is an overall correlation between the CFD-based spatiotemporal unmixedness and the measured emissions. This correlation shows increasing emissions with decreasing mixing quality. Furthermore, the correlation based on the spatiotemporal unmixedness fits the modeled NO_x unmixedness dependency. However, for the spatial unmixedness no correlation can be observed. The measured NO_x emissions would not fit the model anymore when using the spatial definition of the unmixedness. It can be concluded that only a spatiotemporal unmixedness quantity can provide useful results for the cases, considered. The spatiotemporal unmixedness quantity is usually obtained based on transient simulations (e.g. LES or URANS).

For further investigations, the modeled emissions are divided into flame front (below the dashed line) and post flame (above the dashed line) contribution. It can be seen, that the contribution of the flame front NO_x is smaller than the post flame contribution. It can also be observed that the post flame NO_x emissions are more influenced by the unmixedness than the flame front emissions. This can be explained by the increasing contribution of thermal

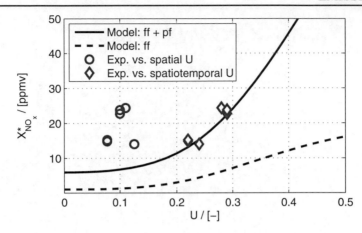

Figure 7.18: Measured NO_x emissions related to the CFD-based unmixedness and comparison against the modeled emissions

NO_x emissions in the post flame. As already seen in Fig. 7.17, below about 10 % unmixedness the main NO_x contribution is based on the residence time effect (i.e. post flame) which is fixed for a given combustion chamber. The overall trend in changing NO_x emissions due to a change in the hardware is predicted very well based on the spatiotemporal definition of the unmixedness.

Chapter 8

Conclusions & summary

8.1 Lessons learned

The objective of this work was to describe and validate methods for the prediction of emissions (NO_x and CO) and thermoacoustics. A low order approach for prediction of the NO_x emissions and a high fidelity CFD-based approach for the combined prediction of emissions and thermoacoustics were presented within this work. The methods were selected and developed based on analysis of the current state of the art. In the next paragraphs the lessons learned, findings and achievements are tied together. Also gaps and limitations are outlined to give recommendations for the next steps.

Within this work a CFD-based approach for the prediction of emissions and thermoacoustics was developed and validated. The proposed method combines a two-zone tabulated chemistry approach, artificial thickening and fully compressible Navier Stokes equations. The fully compressible approach enables acoustic predictions. The applied discretization between first and second order schemes proved to be sufficient to predict intermediate frequency instabilities (e.g. at 250 Hz). The emissions prediction relies on tabulated chemistry, where the post-flame region is identified by CO oxidation. The necessity of a separate post-flame progress variable was pointed out by a part load CO emissions calculation in which the last percentage of the CO oxidation was crucial to distinguish between high and low emissions at the outlet of a combustor. A single whole flame progress variable would not be able to account for those long time-scale effects. Moreover, the prediction of the CO emissions was influenced by transient pressure rate effects which accompanied the initialization. It was shown that this initialization phase might require more computational effort than needed for the averaging of a solution. This should be further investigated and strategies should be established to prevent a very long initialization. The acoustic validation was limited to the prediction of intermediate frequency instabilities. High frequency instabilities above $1000\,Hz$ might require numerical discretization schemes of a higher order for the high fidelity CFD-based approach.

To the knowledge of the author, a combination of thickening and tabulated chemistry has not been used before to predict thermoacoustics with CFD. However, combined prediction, aimed at emissions and acoustics, has also not been shown so far. The literature-based thickening approaches are developed to thicken the turbulent flame, while the proposed method relies on thickening of the laminar prototype flame. The advantage of the novel approach is that the unresolved convection term in the transport equation of the governing species is not affected. Nevertheless, the approach is limited to the flamelet regime. For other combustion regimes at least a validation should be presented. Moreover, the validation was limited to methane type flames and a validation for other fuels, such as hydrogen, might be of interest for future work because those flames might be thinner than methane flames and thickening could be more important.

Within the full-scale validation section an interaction between thermoacoustics effects and emissions was investigated. Thermoacoustically-induced mixture fraction fluctuations were shown to be a reason for a NO_x emissions increase. Therefore, this effect is relevant for the emissions prediction for some acoustically unstable conditions. Nevertheless, the interaction needs further investigation because just one data point was used in this work to show this effect. In future work, the interaction between acoustics, fuel-air unmixedness and emissions might help to obtain more insight into the mechanisms which drive thermoacoustic instabilities.

The importance of transient effects in the context of NO_x emissions was further investigated based on the low order NO_x prediction approach. It could be shown that temporal effects can be crucial in a technically premixed combustion system. It was demonstrated with the full-scale combustion system, that when considering spatiotemporal fluctuations, the calculated unmixedness is about twice as high as the spatial unmixedness without consideration of temporal effects. The spatiotemporal unmixedness correlates perfectly with the proposed low order method for NO_x evaluation. Therefore, the spatiotemporal unmixedness definition should be used for future work.

8.2 Summary

Lean premixed combustion systems have been established as state-of-the-art technology for heavy-duty gas turbines, allowing for low pollutant emissions [5]. However, lean premixed combustion is also associated with thermoacoustic instabilities [6]. Thus, modeling of the key performance parameters – emissions and thermoacoustics – has become mandatory in the design process. The present thesis contributes to the modeling of those key parameters and the validation of the developed methods. The proposed methods can be separated into a low order approach for the NO_x prediction and a high fidelity CFD-based approach for combined prediction of the key performance parameters.

A compressible CFD-based approach, relying on two-zone tabulated chemistry, yields a CFD approach for the prediction of pollutant emissions and thermoacoustics. The addition of a second reaction progress variable enables the emissions prediction. This, in combination with a novel thickening approach, was first applied within this work for a combined emissions and thermoacoustics prediction. The separation between NO_x formation in the flame front and the post-flame region was also applied in a 1D approach in order to provide fast solutions for early design phase calculations and optimization runs.

A stepwise validation of the CFD-based combustion model was presented. First, 1D CFD simulations of freely propagating flames were performed in order to establish the required thickening parameter. Using this parameter, a tolerably small numerical error for the flame propagation speed was achieved. The reference solution of the freely propagating flames was obtained by using detailed chemistry. Subsequently, a lab scale case with GT-relevant conditions was used to calibrate the turbulence chemistry interaction model. The lab scale case was further used in a grid study to determine the required resolution for the technical full-scale validation case.

The novel tabulated chemistry-based CFD combustion model was applied successfully to predict self-excited acoustic instabilities and emissions in a technical GT combustor, using an LES framework. For the prediction of self-excited thermoacoustic instabilities, potential error sources such as boundary condition placement and numerical damping were systematically assessed and excluded. NO_x and CO emissions were predicted at part and base load conditions, while staging and the flow parameter were varied. In consequence, one acoustically unstable case was also part of the investigation. In comparison to the experiments, good qualitative predictions of NO_x emissions were achieved for all conditions. The qualitative CO prediction failed for one of six load points, which could be explained by the fact that the initialization phase of the simulations was still not totally finished. The prediction of intermediate frequency instabilities was achieved for all five acoustically stable cases and the one unstable case.

The low order NO_x emission prediction approach was first validated against perfectly premixed data. Thereafter, a validation at full-scale conditions was performed for the same combustion system as for the CFD-based approach. The experimental reference data could be reproduced with very good qualitative and quantitative accuracy. A spatial and spatiotemporal approach for a CFD-based mixing assessment was compared in the context of the low order approach. Such a CFD-based mixing assessment may serve as an input for the low order tool. Eventually, it was demonstrated that transient effects have to be considered for a CFD-based NO_x assessment.

Using both the 1D approach and the CFD-based approach in combination yielded an improved design process. The low order approach helps in the early design phase to identify core design parameters such as the combustor volume and the required premixing quality. The CFD-based approach, in contrast, improves the evaluation of a draft design with regard to acoustics and emissions. Based on the proposed models, NO_x reduction strategies may be devised. For example, the decision on whether a residence time reduction or an optimization of the premixing performance is required to obtain a given NO_x target can be made based on the low order approach in a first step. In a second step, the premixing passage may be optimized with CFD calculations by using the spatiotemporal unmixedness quantity as a target function for the assessment of the mixing quality. In a final step, the proposed and validated high fidelity CFD approach could be applied to obtain a comprehensive understanding of emissions and thermoacoustics.

In Fig. 8.1, design phases can be distinguished. While the low order approach is dedicated to the early design phase, the right branch can be evaluated by CFD in a later design state. The flow chart illustrates the indirect influence of politics and legislation on the design targets of the customer. The engineer's role is to react and adjust the design from a top level perspective during the whole design process. All decision gates have to be passed to reach the final design, which should be finally tested in experiments. However, it

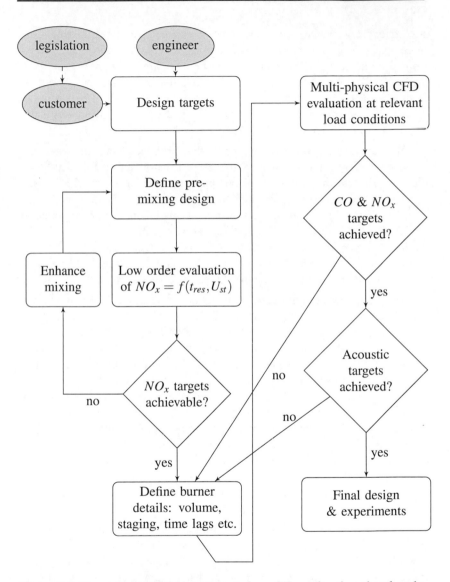

Figure 8.1: Recommended design process for a GT combustion chamber development. The left branch is associated with an early design phase while the right branch aims at a later design phase.

might happen that a target cannot be achieved due to physical or cost reasons. In this case the targets have to be adjusted, which may lead to a limited operational range. All in all, the presented methods can be applied to support the whole design process from an early phase until the final design. The methods are supposed to be predictive, as validated in this work.

For future work, the proposed methods should be validated for further types of combustion systems and other types of fuel. However, the initialization phase of the CFD-based approach also showed some gaps regarding the adjustment of the pressure. It was shown in this work that the mean pressure field might take a long time to become adjusted to the final state. This would lead to an accumulation or reduction of mass in the system during the initialization. As a consequence, the mixture fraction would deviate from the target value, which could affect the CO prediction. To overcome this problem, a computationally cheap initialization strategy based on coarse grids may be established. Another step towards a better assessment of the proposed methods would be the evaluation of the low order and the CFD-based approaches under the same conditions in order to compare their predictability.

Bibliography

[1] Bowman, C. T., 1992. "Control of combustion-generated nitrogen oxide emissions: Technology driven by regulation". *Symposium (International) on Combustion, 24*(1), pp. 859 – 878. Twenty-Fourth Symposium on Combustion.

[2] Raub, J., 1999. "Health effects of exposure to ambient carbon monoxide". *Chemosphere - Global Change Science, 1*(1-3), pp. 331 – 351.

[3] Shindell, D., Faluvegi, G., Stevenson, D., Krol, M., Emmons, L., Lamarque, J.-F., Petron, G., Dentener, F., Ellingsen, K., Schultz, M., et al., 2006. "Multimodel simulations of carbon monoxide: Comparison with observations and projected near-future changes". *Journal of Geophysical Research: Atmospheres (1984–2012), 111*(D19).

[4] Solomon, S., Plattner, G.-K., Knutti, R., and Friedlingstein, P., 2009. "Irreversible climate change due to carbon dioxide emissions". *Proceedings of the national academy of sciences, 106*(6), pp. 1704–1709.

[5] Dederichs, S., Zarzalis, N., Habisreuther, P., Beck, C., Prade, B., and Krebsb, W., 2013. "Assessment of a Gas Turbine NOx Reduction Potential Based on a Spatiotemporal Unmixedness Parameter". *ASME J. Eng. Gas Turbines Power, 135*.

[6] Dowling, A. P., and Stow, S. R., 2003. "Acoustic analysis of gas turbine combustors". *Journal of propulsion and power, 19*(5), pp. 751–764.

[7] Hermeth, S., Staffelbach, G., Gicquel, L. Y., Anisimov, V., Cirigliano, C., and Poinsot, T., 2014. "Bistable swirled flames and influence on

flame transfer functions". *Combustion and Flame,* *161*(1), pp. 184 – 196.

[8] Steele, R. C., Jarrett, A. C., Malte, P. C., Tonouchi, J. H., and Nicol, D. G., 1997. "Variables Affecting NOx Formation in Lean-Premixed Combustion". *ASME J. Eng. Gas Turbines Power,* *119*, pp. 102–107.

[9] Bulat, G., Fedina, E., Fureby, C., Meier, W., and Stopper, U., 2014. "Reacting flow in an industrial gas turbine combustor: Les and experimental analysis". *Proceedings of the Combustion Institute.* In Press, Corrected Proof.

[10] Dederichs, S., Zarzalis, N., and Beck, C., 2015. "Validation of a novel LES approach using tabulated chemistry for thermo-acoustic instability prediction in gas turbines". *Proceedings of ASME,* GT2015-43502.

[11] Poinsot, T., and Veynante, D., 2005. *Theoretical and Numerical Combustion,* 2nd ed. R.T. Edwards.

[12] Kern, M., 2013. "Modellierung kinetisch kontrollierter, turbulenter Flammen für Magerbrennkonzepte". PhD thesis, Karlsruher Institut für Technologie (KIT).

[13] Schmid, H.-P., 1995. "Ein Verbrennungsmodell zur Beschreibung von Wärmefreisetzung von vorgemischten turbulenten Flammen". PhD thesis, Universität Karlsruhe (T.H.).

[14] Hirschfelder, J., Curtiss, C., Bird, R., and of Wisconsin. Theoretical Chemistry Laboratory, U., 1954. *Molecular theory of gases and liquids.* Structure of matter series. Wiley.

[15] Wilke, C. R., 1950. "A Viscosity Equation for Gas Mixtures". *Jourbal of Chemical Physics,* *18*(4), April, pp. 517 – 519.

[16] Bird, R., Stewart, W., and Lightfoot, E., 2013. *Transport Phenomena.* John Wiley & Sons, Limited.

[17] Richardson, L. F., 1922. *Weather prediction by numerical process.* Cambridge University Press.

[18] Kolmogorov, A. N., 1941. "The local structure of turbulence in incompressible viscous fluid for very large Reynolds numbers". In Dokl. Akad. Nauk SSSR, Vol. 30, JSTOR, pp. 301–305.

[19] Borghi, R., 1988. "Turbulent combustion modelling ". *Progress in Energy and Combustion Science, 14*(4), pp. 245 – 292.

[20] Peters, N., 1988. "Laminar flamelet concepts in turbulent combustion". *Symposium (International) on Combustion, 21*(1), pp. 1231 – 1250.

[21] Smith, G. P., Golden, D. M., Frenklach, M., Moriarty, N. W., Eiteneer, B., Goldenberg, M., Bowman, C. T., Hanson, R. K., Song, S., Jr., W. C. G., Lissianski, V. V., and Qin, Z., 1995. Gas Research Institute.

[22] Prade, B., 2013. "Gas turbine operation and combustion performance issues". In *Modern Gas Turbine Systems*, P. Jansohn, ed., Woodhead Publishing Series in Energy. Woodhead Publishing, pp. 383 – 423e.

[23] Zeldovich, Y. B., 1946. "The oxidation of nitrogen in combustion and explosions". *Acta Physicochim. URSS, 21*(4), pp. 577–628.

[24] Lyons, V. J., 1982. "Fue/Air Nonuniformity-Effect on Nitric Oxide Emissions". *AIAA Journal*, pp. 660–665.

[25] Pompei, F., and Heywood, J. B., 1972. "The role of mixing in burner-generated carbon monoxide and nitric oxide". *Combustion and Flame, 19*(3), pp. 407 – 418.

[26] Franzelli, B., Riber, E., Gicquel, L. Y., and Poinsot, T., 2012. "Large Eddy Simulation of combustion instabilities in a lean partially premixed swirled flame". *Combustion and Flame, 159*(2), pp. 621 – 637.

[27] Fenimore, C., 1971. "Formation of nitric oxide in premixed hydrocarbon flames". *Symposium (International) on Combustion, 13*(1), pp. 373 – 380. Thirteenth symposium (International) on Combustion Thirteenth symposium (International) on Combustion.

[28] Malte, P., and Pratt, D., 1975. "Measurement of atomic oxygen and nitrogen oxides in jet-stirred combustion". *Symposium (International) on Combustion, 15*(1), pp. 1061 – 1070. Fifteenth Symposium (International) on Combustion.

[29] Bulat, G., Jones, W., and Marquis, A., 2014. "NO and CO formation in an industrial gas-turbine combustion chamber using LES with the Eulerian sub-grid PDF method". *Combustion and Flame, 161*(7), pp. 1804 – 1825.

[30] Watson, G. M., Munzar, J. D., and Bergthorson, J. M., 2014. "NO formation in model syngas and biogas blends". *Fuel, 124*(0), pp. 113 – 124.

[31] Leonard, G., and Stegmaier, J., 1994. "Development of an aeroderivative gas turbine dry low emissions combustion system". *ASME J. Eng. Gas Turbines Power, 116*(3), pp. 542–546.

[32] Cannon, S., Brewster, B., and Smoot, L., 1998. "Stochastic Modeling of CO and NO in Premixed Methane Combustion". *Combustion and Flame, 113*(1-2), pp. 135 – 146.

[33] Michaud, M. G., Westmoreland, P. R., and Feitelberg, A. S., 1992. "Chemical mechanisms of NOx formation for gas turbine conditions". *Symposium (International) on Combustion, 24*(1), pp. 879 – 887. Twenty-Fourth Symposium on Combustion.

[34] Miller, J. A., and Bowman, C. T., 1989. "Mechanism and modeling of nitrogen chemistry in combustion". *Progress in Energy and Combustion Science, 15*(4), pp. 287–338.

[35] Correa, S., Drake, M., Pitz, R., and Shyy, W., 1985. "Prediction and measurement of a non-equilibrium turbulent diffusion flame". *Symposium (International) on Combustion, 20*(1), pp. 337–343.

[36] Correa, S., and Gulati, A., 1989. "Non-premixed turbulent CO/H2 flames at local extinction conditions". *Symposium (International) on Combustion, 22*(1), pp. 599 – 606.

[37] Correa, S. M., 1992. "Carbon monoxide emissions in lean premixed combustion". *Journal of Propulsion and Power, 8*(6), pp. 1144–1151.

[38] Rijke, P., 1859. "Notiz über eine neue Art, die in einer an beiden Enden offenen Röhre enthaltene Luft in Schwingungen zu versetzen". *Annalen der Physik, 183*(6), pp. 339–343.

[39] Rayleigh, J., 1878. "The explanation of certain acoustical phenomena". *Nature, 18*(455), pp. 319–321.

[40] Hubbard, S., and Dowling, A., 2001. "Acoustic resonances of an industrial gas turbine combustion system". *Journal of engineering for gas turbines and power, 123*(4), pp. 766–773.

[41] Bellows, B. D., Bobba, M. K., Seitzman, J. M., and Lieuwen, T., 2007. "Nonlinear flame transfer function characteristics in a swirl-stabilized combustor". *Journal of Engineering for Gas Turbines and Power, 129*(4), pp. 954–961.

[42] Krediet, H., 2012. "Prediction of limit cycle Pressure Oscillations in Gas Turbine Combustion Systems using the Flame Describing Function". PhD thesis, University of Twente.

[43] Krediet, H., Portillo, J., Krebs, W., and Kok, J., 2010. "Prediction of Thermoacoustic Limit Cycles during Premixed Combustion using the Modified Galerkin Approach". *AIAA/ASME/SAE/ASEE Joint Propulsion Conference and Exhibit, 46*. Nashville.

[44] Dowling, A. P., 1997. "Nonlinear self-excited oscillations of a ducted flame". *Journal of Fluid Mechanics, 346*, pp. 271–290.

[45] Fischer, A., Hirsch, C., and Sattelmayer, T., 2006. "Comparison of multi-microphone transfer matrix measurements with acoustic network models of swirl burners". *Journal of Sound and Vibration, 298*, pp. 73 – 83.

[46] Gelbert, G., Moeck, J. P., Paschereit, C. O., and King, R., 2012. "Feedback control of unstable thermoacoustic modes in an annular Rijke tube". *Control Engineering Practice, 20*(8), pp. 770 – 782.

[47] Preetham, S. H., and Lieuwen, T. C., 2007. "Response of turbulent premixed flames to harmonic acoustic forcing". *Proceedings of the Combustion Institute, 31*(1), pp. 1427 – 1434.

[48] Staffelbach, G., Gicquel, L., Boudier, G., and Poinsot, T., 2009. "Large Eddy Simulation of self excited azimuthal modes in annular combustors". *Proceedings of the Combustion Institute, 32*(2), pp. 2909 – 2916.

[49] Zimont, V., 1999. "Gas premixed combustion at high turbulence. Turbulent flame closure combustion model". *Experimental Thermal and Fluid Science,* *21*(1-3), pp. 179 – 186.

[50] Hawkes, E., and Cant, R., 2001. "Implications of a flame surface density approach to large eddy simulation of premixed turbulent combustion". *Combustion and Flame,* *126*(3), pp. 1617 – 1629.

[51] Ma, T., Stein, O., Chakraborty, N., and Kempf, A., A., 2013. "A posteriori testing of algebraic flame surface density models for LES". *Combustion Theory and Modelling,* *17*(3), pp. 431–482.

[52] Zhang, F., Habisreuther, P., and Bockhorn, H., 2013. "Application of the Unified Turbulent Flame-Speed Closure (UTFC) Combustion Model to Numerical Computation of Turbulent Gas Flames". *High Performance Computing in Science and Engineering,* pp. 187–205.

[53] Pitsch, H., 2005. "A consistent level set formulation for large-eddy simulation of premixed turbulent combustion". *Combustion and Flame,* *143*(4), pp. 587 – 598. Special Issue to Honor Professor Robert W. Bilger on the Occasion of His Seventieth Birthday.

[54] Domenico, M. D., Gerlinger, P., and Noll, B., 2011. "Numerical Simulations of Confined, Turbulent, Lean, Premixed Flames Using a Detailed Chemistry Combustion Model". *Proceedings of ASME,* GT2011-45520, pp. 519–530.

[55] Jones, W., Marquis, A., and Prasad, V., 2012. "LES of a turbulent premixed swirl burner using the Eulerian stochastic field method ". *Combustion and Flame,* *159*(10), pp. 3079 – 3095.

[56] Janicka, J., and Sadiki, A., 2005. "Large eddy simulation of turbulent combustion systems". *Proceedings of the Combustion Institute,* *30*(1), pp. 537 – 547.

[57] van Oijen, J., Lammers, F., and de Goey, L., 2001. "Modeling of complex premixed burner systems by using flamelet-generated manifolds". *Combustion and Flame,* *127*(3), pp. 2124 – 2134.

[58] Ketelheun, A., Olbricht, C., Hahn, F., and Janicka, J., 2011. "NO prediction in turbulent flames using LES/FGM with additional transport

equations". *Proceedings of the Combustion Institute,* *33*(2), pp. 2975 – 2982.

[59] Wetzel, F., Habisreuther, P., and Zarzalis, N., 2006. "Numerical investigation of lean blow out of a model gas turbine combustion chamber using a presumed JPDF-reaction model by taking heat loss processes into account". *Proceedings of ASME,* GT2006-90064, pp. 41–49.

[60] Gicquel, O., Darabiha, N., and Thavenin, D., 2000. "Liminar premixed hydrogen/air counterflow flame simulations using flame prolongation of ILDM with differential diffusion". *Proceedings of the Combustion Institute,* *28*(2), pp. 1901 – 1908.

[61] Fiorina, B., Gicquel, O., Vervisch, L., Carpentier, S., and Darabiha, N., 2005. "Approximating the chemical structure of partially premixed and diffusion counterflow flames using FPI flamelet tabulation". *Combustion and Flame,* *140*(3), pp. 147 – 160.

[62] Kuenne, G., Ketelheun, A., and Janicka, J., 2011. "LES modeling of premixed combustion using a thickened flame approach coupled with FGM tabulated chemistry". *Combustion and Flame,* *158*(9), pp. 1750 – 1767.

[63] Butler, T., and O'Rourke, P., 1977. "A numerical method for two dimensional unsteady reacting flows". *Symposium (International) on Combustion,* *16*(1), pp. 1503 – 1515.

[64] Goodwin, D., Malaya, N., Moffat, H., and Speth, R. Cantera - An objectoriented software toolkit for chemical kinetics, thermodynamics and transport processes. Version 2.0.2.

[65] Konnov, A., 2009. "Implementation of the NCN pathway of prompt-NO formation in the detailed reaction mechanism". *Combustion and Flame,* *156*(11), pp. 2093 – 2105.

[66] Bakali, A. E., Pillier, L., Desgroux, P., Lefort, B., Gasnot, L., Pauwels, J., and da Costa, I., 2006. "NO prediction in natural gas flames using GDF-Kin 3.0 mechanism NCN and HCN contribution to prompt-NO formation". *Fuel,* *85*(7-8), pp. 896 – 909.

[67] Pope, S., 1985. "PDF methods for turbulent reactive flows". *Progress in Energy and Combustion Science,* **11**(2), pp. 119 – 192.

[68] Schmid, H.-P., Habisreuther, P., and Leuckel, W., 1998. "A Model for Calculating Heat Release in Premixed Turbulent Flames". *Combustion and Flame,* **113**, pp. 79 – 91.

[69] Proch, F., and Kempf, A. M., 2014. "Numerical analysis of the Cambridge stratified flame series using artificial thickened flame LES with tabulated premixed flame chemistry". *Combustion and Flame,* **161**(10), pp. 2627 – 2646.

[70] Wegner, B., Gruschka, U., Krebs, W., Egorov, Y., Forkel, H., Ferreira, J., and Aschmoneit, K., 2011. "CFD Prediction of Partload CO Emissions Using a Two-Timescale Combustion Model". *ASME J. Eng. Gas Turbines Power,* **133**(7).

[71] Fric, T. F., 1993. "Effects of fuel-air unmixedness on NOx emissions". *Journal of Propulsion and Power,* **9**, pp. 708–713.

[72] Biagioli, F., and Guethe, F., 2007. "Effect of pressure and fuel air unmixedness on NOx emissions from industrial gas turbine burners". *Combustion and Flame,* **151**, pp. 274 – 288.

[73] Prade, B., Streb, H., Berenbrink, P., Schetter, B., and Pyka, G., 1996. "Development of an Improved Hybrid Burner - Initial Operating Experience in a Gas Turbine". *ASME Paper*(96-GT), p. 45.

[74] Fichet, V., Kanniche, M., Plion, P., and Gicquel, O., 2010. "A reactor network model for predicting NOx emissions in gas turbines". *Fuel,* **89**(9), pp. 2202 – 2210.

[75] Holdeman, J. D., Liscinsky, D. S., and Bain, D. B., 1999. "Mixing of multiple jets with a confined subsonic crossflow: Part II - opposed rows of orifices in rectangular ducts". *ASME J. Eng. Gas Turbines Power,* **121**(3), pp. 551–562.

[76] Schneiders, T., Hoeren, A., Michalski, B., Pfost, H., Scherer, V., and Koestlin, B., 2001. "Investigation of unsteady gas mixing processes in gas turbine burners applying a tracer-LIF method". In ASME Turbo

Expo 2001: Power for Land, Sea, and Air, American Society of Mechanical Engineers, pp. V002T02A016–V002T02A016.

[77] Galeazzo, F. C. C., Donnert, G., Habisreuther, P., Zarzalis, N., Valdes, R. J., and Krebs, W., 2011. "Measurement and simulation of turbulent mixing in a jet in crossflow". *ASME J. Eng. Gas Turbines Power,* *133*(6).

[78] Ivanova, E., Di Domenico, M., Noll, B., and Aigner, M., 2009. "Unsteady simulations of flow field and scalar mixing in transverse jets". *Proceedings of ASME*, GT2009-59147, pp. 101–110.

[79] Ivanova, E., Noll, B., and Aigner, M., 2010. "Unsteady simulations of turbulent mixing in jet in crossflow". *AIAA Fluid Dynamics Conference and Exhibit,* *40*.

[80] Syed, K. J., Roden, K., and Martin, P., 2007. "A novel approach to predicting NOx emissions from dry low emissions gas turbines". *ASME J. Eng. Gas Turbines Power,* *129*(3), pp. 672–679.

[81] Lacarelle, A., Goeke, S., and Paschereit, C. O., 2010. "A quantitative link between cold-flow scalar unmixedness and NOx emissions in a conical premixed burner". *Proceedings of ASME*, GT2010-23132, pp. 919–931.

[82] Cha, C., and Kramlich, J., 2000. "Modeling finite-rate mixing effects in reburning using a simple mixing model". *Combustion and Flame,* *122*(12), pp. 151 – 164.

[83] Correa, S., and Braaten, M., 1993. "Parallel simulations of partially stirred methane combustion". *Combustion and Flame,* *94*(4), pp. 469 – 486.

[84] Ihme, M., and Pitsch, H., 2008. "Modeling of radiation and nitric oxide formation in turbulent nonpremixed flames using a flamelet/progress variable formulation". *Physics of Fluids,* *20*(5), p. 055110.

[85] OPENFOAM FOUNDATION, 2010. *CFD software package: Open-FOAM*. Version 1.7.x.

[86] Patankar, S., 1980. *Numerical Heat Transfer and Fluid Flow*. Series in computational methods in mechanics and thermal sciences. Taylor & Francis.

[87] Pope, S. B., 1999. "A perspective on turbulence modeling". In *Modeling Complex Turbulent Flows*. Springer, pp. 53–67.

[88] Boussinesq, J., 1877. *Essai sur la théorie des eaux courantes*. Imprimerie nationale.

[89] Smagorinsky, J., 1963. "General Circulation Experiments with the primitive Equations". *American Meteorological Society*.

[90] Yoshizawa, A., and Horiuti, K., 1985. "A statistically-derived subgrid-scale kinetic energy model for the large-eddy simulation of turbulent flows". *Journal of the Physical Society of Japan*, *54*(8), pp. 2834–2839.

[91] Kim, W.-W., and Menon, S., 1995. "A new dynamic one-equation subgrid-scale model for large eddy simulations". *American Institute of Aeronautics and Astronautics*, *33*.

[92] Sohankar, A., Davidson, L., and Norberg, C., 1999. "Large Eddy Simulation of Flow Past a Square Cylinder: Comparison of Different Subgrid Scale Models". *Journal of Fluids Engineering*, *122*, pp. 39 – 47.

[93] Fureby, C., Tabor, G., Weller, H., and Gosman, A., 1997. "Differential subgrid stress models in large eddy simulations". *Physics of Fluids*, *9*(11), pp. 3578–3580.

[94] Fureby, C., Alin, N., Wikström, N., Menon, S., Svanstedt, N., and Persson, L., 2004. "Large eddy simulation of high-Reynolds-number wall bounded flows". *AIAA journal*, *42*(3), pp. 457–468.

[95] Schumann, U., 1975. "Subgrid scale model for finite difference simulations of turbulent flows in plane channels and annuli". *Journal of computational physics*, *18*(4), pp. 376–404.

[96] Pope, S. B., 2000. *Turbulent flows*. Cambridge university press.

[97] Crank, J., and Nicolson, P., 1947. "A practical method for numerical evaluation of solutions of partial differential equations of the heat-conduction type". pp. 50–67.

[98] Jasak, H., 1996. "Error analysis and estimation for the finite volume method with applications to fluid flows". PhD thesis, Imperial College London (University of London).

[99] Klein, M., Sadiki, A., and Janicka, J., 2003. "A digital filter based generation of inflow data for spatially developing direct numerical or large eddy simulations". *Journal of Computational Physics,* *186*(2), pp. 652 – 665.

[100] Weiß, M., Zarzalis, N., and Suntz, R., 2008. "Experimental study of Markstein number effects on laminar flamelet velocity in turbulent premixed flames". *Combustion and Flame,* *154*(4), pp. 671 – 691.

[101] Proch, F., and Kempf, A., 2014. "Modeling heat loss effects in the large eddy simulation of a model gas turbine combustor with premixed flamelet generated manifolds". *Proceedings of the Combustion Institute*. In Press, Corrected Proof.

[102] Fiorina, B., Vicquelin, R., Auzillon, P., Darabiha, N., Gicquel, O., and Veynante, D., 2010. "A filtered tabulated chemistry model for LES of premixed combustion". *Combustion and Flame,* *157*(3), pp. 465 – 475.

[103] Pope, S., 2000. *Turbulent Flows*. Cambridge University Press.

[104] Burcat, A., and , B. J. M., 1994. *1994 Ideal Gas Thermodynamic Data for Combustion and Air Pollution Use*. Technion-IIT, Faculty of Aerospace Engineering.

[105] Damköhler, G., 1940. "Der einfluß der Turbulenz auf die Flammengeschwindigkeit in Gasgemischen". *Zeitschrift für Elektrochemie und angewandte physikalische Chemie,* *46*(11), pp. 601–626.

[106] Bradley, D., 1992. "How fast can we burn?". *Symposium (International) on Combustion,* *24*(1), pp. 247 – 262.

[107] Driscoll, J. F., 2008. "Turbulent premixed combustion: Flamelet structure and its effect on turbulent burning velocities". *Progress in Energy and Combustion Science,* **34**(1), pp. 91 – 134.

[108] Durand, L., and Polifke, W., 2007. "Implementation of the Thickened Flame Model for Large Eddy Simulation of Turbulent Premixed Combustion in a Commercial Solver". *Proceedings of ASME,* GT2007-28188, pp. 869–878.

[109] Kuenne, G., Seffrin, F., Fuest, F., Stahler, T., Ketelheun, A., Geyer, D., Janicka, J., and Dreizler, A., 2012. "Experimental and numerical analysis of a lean premixed stratified burner using 1D Raman/Rayleigh scattering and large eddy simulation". *Combustion and Flame,* **159**(8), pp. 2669 – 2689.

[110] Mongia, R. K., Tomita, E., Hsu, F. K., Talbot, L., and Dibble, R. W., 1996. "Use of an optical probe for time-resolved in situ measurement of local air-to-fuel ratio and extent of fuel mixing with applications to low NOx emissions in premixed gas turbines". *Symposium (International) on Combustion,* **26**(2), pp. 2749 – 2755.

[111] Burgess, D. S., 1962. *Structure and propagation of turbulent Bunsen flames.* US Government Printing Office.

[112] Ghirelli, F., and Leckner, B., 2004. "Transport equation for the local residence time of a fluid". *Chemical Engineering Science,* **59**(3), pp. 513 – 523.

[113] Habisreuther, P., 2002. "Untersuchung zur bildung von thermischem stickoxid in turbulenten drallflammen". PhD thesis, Universität Karlsruhe (TH).

[114] Forney, L., and Nafia, N., 1998. "Turbulent Jet Reactors: Mixing Time Scales". *Chemical Engineering Research and Design,* **76**(6), pp. 728 – 736. Process Design.

[115] Lammel, O., Roediger, T., Stoehr, M., Ax, H., Kutne, P., Severin, M., Griebel, P., and Aigner, M., 2014. "Investigation of Flame Stabilization in a High-Pressure Multi-Jet Combustor by Laser Measurement Techniques". *Proceedings of ASME,* GT2014-26376.

[116] de Villiers, E., 2006. "The Potential of Large Eddy Simulation for the Modeling of Wall Bounded Flows". PhD thesis, Imperial College of Science, London.

[117] Guelder, O. L., and Smallwood, G. J., 1995. "Inner cutoff scale of flame surface wrinkling in turbulent premixed flames". *Combustion and Flame,* *103*(1-2), pp. 107 – 114.

[118] Carroni, R., Griffin, T., Mantzaras, J., and Reinke, M., 2003. "High-pressure experiments and modeling of methane/air catalytic combustion for power-generation applications ". *Catalysis Today,* *83*, pp. 157 – 170. 5th International Workshop on Catalytic Combustion.

[119] Beck, C., Braun, S., Breitenbücher, J., Dederichs, S., Deiss, O., Grieb, T., Kadji, A., Schneider, O., and Vogtmann, D., 2013. Brenneranordnung für eine Gasturbine, May 29. Patent EP2597374 A1.

[120] Moase, W. H., Brear, M. J., and Manzie, C., 2004. "Linear and Non-linear Acoustic Behaviour of Outlet Nozzles". *Proceedings of Australasian Fluid Mechanics Conference,* *15*.

[121] Beck, C. H., 2009. "Analyse der Stickoxidbildung in mageren Sprayflammen mit partieller Vorverdunstung". PhD thesis, Karlsruhe Institute of Technology.

[122] Goevert, S., 2011. "Analysis of Unsteady Flow Phenomena in Gas Turbine Combustion Systems Based on Large Eddy Simulations". Master's thesis, University Magdeburg.

[123] Bronstein, I. N., Hromkovic, J., Luderer, B., Schwarz, H.-R., Blath, J., Schied, A., Dempe, S., Wanka, G., Gottwald, S., Zeidler, E., et al., 2012. *Taschenbuch der Mathematik*, Vol. 1. Springer.